CRC Handbook of Plant Cytochemistry

Volume I
Cytochemical Localization of Enzymes

Editor
Kevin C. Vaughn

Plant Physiologist
Weed Biology and Mechanisms of Control Research
Southern Weed Science Laboratory
USDA-Agriculture Research Service
Stoneville, Mississippi

CRC Press, Inc.
Boca Raton, Florida

Library of Congress Cataloging-in-Publication Data

Handbook of plant cytochemistry.

  Bibliography: p.
  Includes index.
  1. Plant cytochemistry—Handbooks, manuals, etc.
I. Vaughn, Kevin C.
QK861.H34   1987   581.87'6042   86-18772
ISBN 0-8493-3247-8 (set)
ISBN 0-8493-3248-6 (vol. I)
ISBN 0-8493-3249-4 (vol. II)

This book represents information obtained from authentic and highly regarded sources. Reprinted material is quoted with permission, and sources are indicated. A wide variety of references are listed. Every reasonable effort has been made to give reliable data and information, but the author and the publisher cannot assume responsibility for the validity of all materials or for the consequences of their use.

All rights reserved. This book, or any parts thereof, may not be reproduced in any form without written consent from the publisher.

Direct all inquiries to CRC Press, Inc., 2000 Corporate Blvd., N.W., Boca Raton, Florida, 33431.

© 1987 by CRC Press, Inc.

International Standard Book Number 0-8493-3247-8 (set)
International Standard Book Number 0-8493-3248-6 (v. 1)
International Standard Book Number 0-8493-3249-4 (v. 2)

Library of Congress Card Number 86-18772
Printed in the United States

PREFACE

Plant cytochemistry, as well as most areas of plant sciences, has relied heavily on the procedures from zoological/medical research. Only the procedures for malate synthase developed by Trelease and colleagues (see chapter entitled "Malate Synthase") has followed the opposite route of a plant technique being used to detect an animal enzyme. Even the procedures for the detection of photosynthetic electron transport, an area sacred and unique to plants, was developed originally by animal workers at the National Institute of Health! Even with the extensive borrowing of animal techniques that has gone on, plants present some unique problems for cytochemists that have not brought ready solutions from animal procedures. Being bound by cell walls, plant cells have an additional barrier for reagents to penetrate not encountered by animal cells. This represents special problems for immunocytochemistry and lectin studies where antibodies or lectins must be allowed to enter through cut ends of cells or protoplasts must be made permeable to them by fixation or permeabilization with detergents for pre-embedding procedures. The extensive development of postembedding staining procedures (again mainly borrowed from zoological techniques!) promises to open up these procedures to more plant questions. The localization of the various photosystem complexes in the chloroplast using postembedding procedures is an indication of more exciting developments in this area.

THE EDITOR

Kevin C. Vaughn, Ph.D., is a Research Plant Physiologist in the Southern Weed Science Laboratory of Agricultural Research Service, U.S. Department of Agriculture, Stoneville, Mississippi.

He received a B.A. degree with honors in biology from Clark University in 1976 and a Ph.D. in Botany with emphasis in genetics from Miami University (Ohio) in 1980. Since that time, he has been employed by the Southern Weed Science Laboratory working on the mode of action of the fungal toxin tentoxin and on the mechanisms for herbicide resistance in various weed biotypes.

He is a member of the American Society of Plant Physiologists, the Japanese Society of Plant Physiologists, the Scandinavian Society of Plant Physiologists, the American Genetics Association, the Botanical Society of America, and the Weed Science Society of America. He serves on the editorial board of *Plant Physiology*.

Dr. Vaughn has over 70 publications in the areas of plant cell biology and cytochemistry. Dr. Vaughn was USDA's Mid-South Area Scientist of the Year in 1985.

Aside from his professional interest in plants, Dr. Vaughn has hybridized plants since he was 10 and has a number of award-winning cultivars of hosta, iris, and sempervivum to his credit.

ADVISORY BOARD

Sue Ellen Fredericks
Department of Biological Sciences
Mount Holyoke College
South Hadley, Massachusetts

Peter Gruber
Department of Biological Sciences
Mount Holyoke College
South Hadley, Massachusetts

Martha J. Powell
Department of Botany
Miami University
Oxford, Ohio

John M. Robinson
Department of Pathology
Harvard Medical School
Boston, Massachusetts

Kenneth G. Wilson
Department of Botany
Miami University
Oxford, Ohio

CONTRIBUTORS

Randy D. Allen
Research Assistant
Department of Biology
Texas A & M University
College Station, Texas

David H. Clapham
Research Assistant
Department of Plant and Forest Protection
Swedish University of Agricultural Sciences
Uppsala, Sweden

Mihály Ekés
Department of Plant Anatomy
Eötvös Loránd University
Budapest, Hungary

Sue Ellen Frederick
Professor
Department of Biological Sciences
Mount Holyoke College
South Hadley, Massachusetts

Albert P. Kausch
Department of Cell Biology
The Rockefeller University
New York, New York

C. Edward McClelen
Research Associate
Department of Biology
Baylor University
Waco, Texas

Randy Moore
Associate Professor of Biology
Department of Biology
Baylor University
Waco, Texas

Craig L. Nessler
Associate Professor
Department of Biology
Texas A & M University
College Station, Texas

Maria Salomé Soares Pais
Full Professor
Department of Biologia Vegetal
Faculdade de Ciências da Universidade de Lisboa
Lisbon, Portugal

Rex Paul
Biologist
Department of Weed Biology and Mechanisms of Control Research
Southern Weed Science Laboratory
Stoneville, Mississippi

Houston S. Smith
Research Assistant
Department of Biology
Baylor University
Waco, Texas

Richard N. Trelease
Professor of Biology
Department of Botany and Microbiology
Arizona State University
Tempe, Arizona

Kevin C. Vaughn
Plant Physiologist
Department of Weed Biology and Mechanisms of Control Research
Southern Weed Science Laboratory
USDA-Agricultural Research Service
Stoneville, Mississippi

TABLE OF CONTENTS

Volume I

GENERAL PROCEDURES
DAB Procedures 3
Cerium Precipitate 25
Phosphatases 37
Dehydrogenases 65

SPECIFIC ENZYME PROTOCOLS
Aspartate Aminotransferase 95
Enzyme Cytochemistry and Immunocytochemistry of Nucleases 107
Cytochemical Localization of Lipases in Plant Cells 123
Malate Synthase 133
Pectinase 149
Polyphenol Oxidase 159

INDEX 165

Volume II

Cutinized and Suberized Cell Walls 1
Photosynthetic Partial Reactions 37
Lectins 45
Immunochemical Analysis of Plant Tissue 65
Sodium and Potassium 121
Calcium 155

INDEX 171

General Procedures

DAB PROCEDURES

Sue Ellen Frederick

THE CYTOCHEMISTRY OF DIAMINOBENZIDINE

Diaminobenzidine (DAB) has been among the most widely exploited cytochemical compounds since its debut almost 20 years ago as a detector of horseradish peroxidase (HRP).[1] Its wide application to problems in diverse groups of organisms reflects its extreme flexibility: by relatively subtle changes in conditions under which it is employed, DAB can be used to localize many different heme compounds and enzymes, as well as several electron transport assemblies. The versatility is not an unmixed blessing, however, since it leads directly to the major concern in DAB cytochemistry — the lack of specificity. As this chapter will point out, reliable specificity is attainable in most cases, but only with the judicious use and interpretation of controls.

The roles of DAB in plant cytochemistry are, perhaps, not as far-flung as those in animal cytochemistry. There it has served widely, not only for demonstrating endogenous peroxidases, but also for tracking exogenous peroxidase used to delineate compartments or to map neuronal continuities. Nevertheless, the contributions of DAB to plant biology have been significant and numerous: they range from affording the most convenient means to identify beyond doubt the morphologically variable peroxisome to allowing the visualization of photosystem I activity in chloroplasts. Furthermore, its ability to detect peroxidase has been of paramount value in immunocytochemical work, a topic covered in another chapter of this volume.

Applications and contributions of DAB to various lines of research with plants and fungi will be emphasized here, along with the practical aspects of the procedures. Coverage of the plant literature will be selective; it is intended to provide examples of the topics considered rather than to be exhaustive. Alternative reviews of at least certain areas may be found in a number of other sources.[2-4] As plant DAB cytochemistry is heavily indebted to research with animal cells — not only for the original designs of procedures, but also for continued innovations and refinements — the author has chosen to cite freely, whenever relevant, references to animal work. For a general perspective on the utility of DAB in animal research, the reader is referred to reviews by Essner[5] and Litwin.[6]

DAB REACTIONS OF CYTOCHEMICAL INTEREST

DAB, or 3,3′,4,4′-tetraaminobiphenyl (Figure 1), is a diamino derivative of benzidine, a dye introduced early in the century as a substrate for peroxidase demonstration. The advantage of diaminobenzidine over benzidine is that upon oxidation, either by peroxidatic activity or other agents, it yields a brownish polymer which can be rendered electron dense upon reaction with osmium (Figure 1). In contrast, oxidation of benzidine produces a blue product which is neither osmiophilic nor itself of substantial electron density; its detection is, thus, largely restricted to light microscopy.

Studies of the oxidation of DAB in vitro indicate that polymerization and cyclization (Figure 1) occur during this step.[7] The oxidized polymer is highly insoluble in water as well as common organic solvents, and has a melting point in excess of 300°C. Because it is extremely insoluble, oxidized DAB is assumed to remain close to the site of its formation. Furthermore, while alone it is already somewhat electron dense, the product ("osmium black") which it forms upon treatment with osmium tetroxide is extremely opaque. All these properties render DAB an excellent substrate for cytochemical localizations.

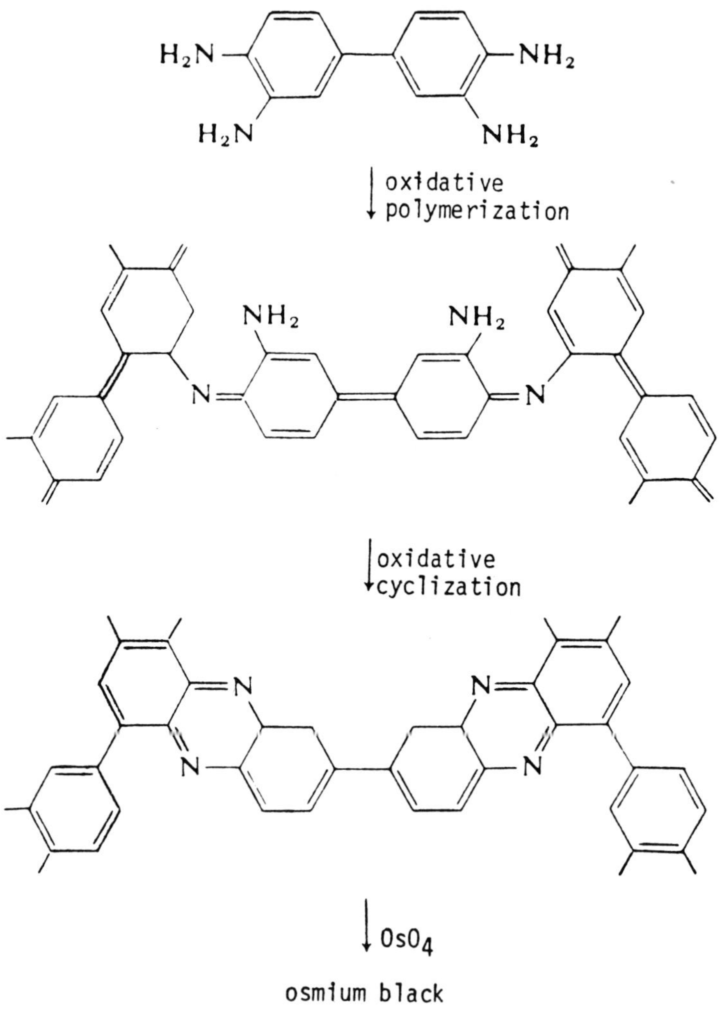

FIGURE 1. Steps leading to "osmium black" formation in DAB cytochemistry, as proposed by Seligman et al.[7] DAB, shown at the top, is first oxidized to a polymeric form (a), which then undergoes cyclization to a complex molecule (b) which is highly osmiophilic.

LOCALIZATION OF CATALASE AND PEROXIDASE

Perhaps the major applications of DAB to plant cytochemistry have been for detection of the heme proteins catalase and peroxidase. As DAB lacks an inherent selectivity for specific heme proteins, a means for distinguishing between these two enzymes must be introduced by careful design of various experimental parameters. The most important of these include the nature of the prefixation, the conditions of the cytochemical incubation, the composition and pH of the incubation medium, and the choice of proper controls. While certain generalizations regarding these parameters have seemed obvious from the beginning, it is now apparent that there is considerable variability among cell types and organisms as to the behavior of both catalase and peroxidase. It is, thus, prudent to preface a discussion of the more or less "standard" procedures with the caveat that for precise work, experiments must be carried out for each type of sample to delineate conditions optimal for revealing a particular hemoprotein activity and to confirm the specificity of the reaction.

Optimizing conditions for cytochemical oxidation of DAB can be facilitated by biochemical assays in vitro, for example, by the spectrophotometric technique of Fahimi and Herzog.[8] Their method quantifies enzyme-mediated oxidation of DAB based on an increase in absorption at 456 nm, a wavelength within the maximum absorption band (455 to 470 nm) previously reported for oxidized DAB products.[9] While parallel assays of the activity of particular enzymes in vitro and observations on their cytochemical reactivity *in situ* have agreed closely in some cases,[10] in at least a few others in vitro results have not forecast which conditions (e.g., pH and concentrations of reagents or inhibitors) are most suitable for cytochemical protocols. In the final analysis, then, the procedures designed for catalase and peroxidase are not infallible, and their potential limitations should be borne in mind.

Catalase

Detection of catalase via DAB cytochemistry depends upon the "peroxidatic" (Equation 1) rather than "catalatic" (Equation 2) activity of this enzyme. In Equation 1, "A" represents DAB or another organic compound used as substrate.

$$AH_2 + H_2O_2 \xrightarrow{\text{catalase}} A + 2H_2O \qquad (1)$$

$$2H_2O_2 \xrightarrow{\text{catalase}} 2H_2O + O_2 \qquad (2)$$

Conditions which optimize peroxidatic activity of catalase and at the same time minimize the activity of true peroxidases should be chosen.

Prefixation

The conditions of the prefixation appear to have a pronounced effect on the occurrence and intensity of the subsequent staining reaction. Fixation with glutaraldehyde has been reported to augment greatly the peroxidatic activity of catalase, at least from mammalian sources, and simultaneously to inhibit its catalatic activity;[11] indeed, unfixed rat liver peroxisomal fractions show no peroxidatic action on DAB. While no similar claims have been established for plant catalase, it might be predicted that glutaraldehyde fixation (or some other treatment which enhances peroxidatic activity) is prerequisite for DAB staining here, also. At the same time that glutaraldehyde fixation seems to elicit the peroxidatic activity of catalase, it also appears that mild conditions of fixation yield the most intense staining. One group[10] reported that the lowest temperature (5°C) and aldehyde concentration (1%) tried for the shortest period of time (30 min) gave the most peroxidatic activity in spectrophotometric assays of DAB oxidation by catalase, as well as in cytochemical tests. There are cases, however, in which higher than optimal glutaraldehyde concentrations may be preferred — for example, when one wishes to suppress the activity of peroxidase or to fix a particularly compact tissue rapidly.[10]

Cytochemical Treatment

Procedures employed for catalase of animals, plants, fungi, and a variety of protists are basically similar to one another and are, for the most part, patterned after the sequence described by Novikoff and Goldfischer[12] for demonstrating catalase in liver peroxisomes. In the usual protocols, fixation of materials in glutaraldehyde is followed first by rinsing and then by incubating in an alkaline solution containing DAB and H_2O_2. A representative protocol including a typical incubation medium is summarized in Table 1.

A central parameter of the cytochemical reaction is the pH of the DAB incubation medium. While values as low as 6.0 give dense peroxisomal staining in some organisms,[13] ordinarily the intensity is greatly increased under alkaline conditions. Furthermore, the competing

Table 1
REPRESENTATIVE PROTOCOLS FOR VISUALIZING CATALASE AND PEROXIDASE WITH DAB

Step	Catalase, tobacco leaves[17]	Peroxidase, wound vessel members of *Coleus*[57]
Prefixation	3% Glutaraldehyde in 0.05 M K-phosphate, pH 6.8; 1.5—2.0 hr; room temperature	2.5% Glutaraldehyde in 0.05 M phosphate, pH 6.8; 1.0—2.0 hr; room temperature
Rinse	0.05 M K-phosphate, pH 6.8; 15—20 min	0.05 M phosphate, pH 6.8; stored in refrigerator up to a week
Incubation Medium	2 mg/mℓ DAB 0.05 M Propanediol, pH 10.0; 0.06% H_2O_2, pH adjusted to 9.0	0.5 mg/mℓ DAB 0.05 M Tris, pH 7.6; 0.01% H_2O_2
Incubation Conditions	50—60 min 37°C	15—60 min Room temperature
Rinse	0.05 M K-phosphate, pH 6.8; 15—20 min	Not specified
Postfixation	2% OsO_4 in 0.05 M K-phosphate; 2 hr	2% OsO_4 (buffer not specified); 2 hr
Controls	DAB medium minus H_2O_2 Preincubation (20 to 30 min) and incubation in 0.01 or 0.001 M KCN Preincubation (20 min) and incubation in 0.02 M AT Anaerobic incubation in DAB medium with and without H_2O_2	DAB medium minus H_2O_2 Preincubation (30 min) and incubation in 0.01 M KCN Preincubation (30 min) and incubation in 0.02 M AT 2 mM Na-pyruvate in DAB medium minus H_2O_2
Viewing of sections	Unstained, or stained with lead citrate and uranyl acetate	Without counterstain

staining due to peroxidase, whose pH optimum is more in the neutral range, is largely eliminated at a high pH. These observations are consistent with the report[14] that glutaraldehyde-treated isolated beef liver catalase shows optimal DAB oxidation at pH 10.5 in spectrophotometric assays. One speculation is that high pH causes dissociation of catalase into peroxidatically active subunits.[15] The buffers actually used to achieve the recommended alkaline ranges are most commonly propanediol, Tris, and glycine.

The concentration of DAB employed is most often between 0.5 and 2 mg/mℓ, while that of H_2O_2 varies more widely (a range of 0.005 to 1.5%). The two are likely related, higher concentrations of DAB requiring higher H_2O_2 concentrations for maximal staining intensity.[10] Some media have included $MnCl_2$, but this addition is not recommended, as DAB can be oxidized by manganese in the absence of enzymatic activity.[16]

Incubation times and temperatures vary somewhat and should be worked out for optimal staining intensity of particular cells/tissues. Staining of plant peroxisomes has been most successful at 37°C and this temperature has been employed in almost all reports. Staining at room temperature is often reduced or absent (e.g., see Frederick and Newcomb[17]), perhaps because of a decreased penetration of the reagents. Some investigators of animal cells have enhanced peroxisomal catalase staining by raising the incubation temperature from 37 to 45°C.[18] Conceivably, such an increase might intensify peroxisomal staining in plants, also, though such results have not yet been reported. Optimal duration of incubation is related to the temperature and, also, to the difficulty of penetration of samples. For tissue blocks or sections, 60 min is most frequently employed, although times from 30 to 90 min are also common. It is suggested that all incubations, and, in particular, lengthy ones, be carried out in the dark, to prevent light-mediated oxidation of the DAB (see later section, "Binding of DAB or DAB Oxides to Tissue Components").

Since DAB penetrates tissues relatively slowly, it may be necessary to use tissue sections rather than blocks for uniform results, or to enhance penetration by some means, such as removal of the lower epidermis of leaves.[17] When tissue blocks are employed, the reaction product may be limited to a peripheral layer. Penetration of reagents must be constantly surveyed in DAB cytochemistry, and it seems prudent to monitor routinely the progress of inward movement by observing light microscope sections cut at all levels of the treated sample. The depth to which the brown oxidized DAB polymer can be visualized serves as an approximate gauge of the degree of reagent penetration; segments to be observed in the electron microscope can then be chosen accordingly.

A variation used for isolated organelle fractions is to preincubate first in a medium containing DAB but no H_2O_2, then to introduce H_2O_2 gradually through dialysis.[19] This procedure, designed to show matrix staining for catalase in isolated glyoxysomes which otherwise stain only at the membrane, is based on the premise that reactions at the periphery of isolated organelles containing catalase may rapidly build up a layer of polymerized product which prevents further access of reagents to the matrix. While this approach has yielded successful matrix staining for catalase, it has not been confirmed that this explanation accounts for it.

Controls

To assume that all staining which occurs during incubation in media designed for optimal visualization of catalase results from activity of that enzyme is not justified. Controls must, therefore, be carried out which distinguish among various possible mediators of the reaction. In particular, it may prove difficult to distinguish between the peroxidatic activity of catalase and the activity of true peroxidases, which it closely mimics. Foremost among the controls is use of 3-amino-1,2,4-triazole (AT), a potent inhibitor of catalase in many plant and animal tissues.[20,21] This compound, often employed at a concentration of $0.02\ M$, totally eliminates staining due to catalase in most plant and animal cell types. It cannot be taken for granted, however, that this agent will always inhibit catalase-mediated DAB staining completely; indeed, catalase activity of maize scutellum, measured spectrophotometrically, was reported to be inhibited only 60 to 70% by $0.02\ M$ AT.[22] This figure may bear little relevance to DAB staining, however, as the cytochemical staining is mediated by the peroxidatic rather than the catalatic reaction of catalase. It is also important not to assume that AT has no inhibitory effect besides that on catalase; in fact, it is known to inhibit peroxidase, also, in some animal tissues (cf. Strum and Karnovsky[23] and Cavallo,[24] for example). Its inhibition of other enzymes is rarely as complete as that of catalase, however, at least in plants. Thus, a positive reaction in the presence of AT would contraindicate activity solely of catalase. It is important that tissues be pretreated for at least 30 min with AT prior to incubation in complete cytochemical medium, in order to allow adequate penetration and reaction of the inhibitor. The compound is also included, at the same concentration, in the incubation medium itself and in the rinses following.

Potassium cyanide (KCN), which should inhibit both catalase and peroxidase,[25] is a less reliable control. It generally fails to eliminate totally the DAB staining of peroxisomes (at concentrations of 0.01 to 0.001 M) in both animal and plant cells.[17,26] On the other hand, KCN does, in some cases, eliminate DAB reactions mediated by peroxidase (see below), and so can be useful for distinguishing between these two activities. As is the case for AT, KCN should be exposed to tissues in a preincubation step (usually 20 to 30 min), then added to the incubation medium and the subsequent rinses.

Catalase-mediated DAB staining should be greatly reduced or eliminated by omission of H_2O_2 from the medium; one must take care to arrest the production of endogenous peroxides for this measure to be effective, however. This can be attempted by one of the following means: carrying out the reaction anaerobically, adding 2 mM sodium pyruvate to the preincubation and incubation medium, or including beef liver catalase (0.5 mg/mℓ) in the incubation medium.[26]

Applications of the Procedure
Identification of Peroxisomes in Higher Plants

Diaminobenzidine has remained the major cytochemical reagent for revealing catalase localization. As this enzyme is the marker for the peroxisome, a prominent and metabolically important organelle but one lacking definitive structural characteristics, DAB has proved indispensable in the identification of this organelle. First used in plants to show equivalence between peroxisomes and the morphologically defined microbodies in leaves,[17] endosperm, and/or other plant parts,[27,28] this reagent has now been applied to all higher plant organs and literally dozens of cell types and species. The following correspondence has now proved the rule among the higher plants, namely, that virtually all microbody-appearing organelles of higher plants show DAB staining, though occasionally the actual pattern of staining exhibited by them may be unusual. This cytochemical identification of peroxisomes has been a valuable asset to plant cell biology generally, as these organelles cannot be isolated from many cell types due to inadequate homogenization techniques or inability to separate a particular cell type from a heterogeneous sample.

DAB staining attributable to catalase is confined in plants to peroxisomes, where it occurs throughout the matrix (Figure 2). (When inclusions are present, the crystals, fibers, or amorphous condensations have all been noted to show pronounced DAB staining, indicative, likely, of concentrated catalase activity.[17,27,29,30] Figure 3 depicts such a DAB-stained inclusion.) While a uniform staining of the peroxisome matrix has been the result in almost every instance, exceptions have been reported. For example, microbodies in spadix appendices of *Arum maculatum,* both isolated and *in situ,* show only membrane staining after procedures for catalase localization are followed.[31] A similar membrane staining of glyoxysomes isolated from maize scutellum has been reported[22] and attributed to a high concentration of membrane-associated catalase. In this case, most of the catalase is known to be present in the matrix, however, and the failure of the DAB procedure to reveal it indicates a possible shortcoming in the technique. As mentioned earlier, rapid DAB polymerization at the periphery of isolated organelles may effectively prevent this substrate from reaching the matrix or, alternatively, the matrix enzyme may be lost from the organelles during isolation. With the *Arum* microbodies, even the membrane staining disappears after anthesis, although catalase-containing microbodies can still be isolated at this stage. Such results point out the need for cautious interpretation of cytochemical findings and suggest that more variability may occur in the reactivity of higher plant catalases toward DAB than is commonly assumed.

Staining of the cytosol is not usually noted after DAB incubation; if it should occur, however, a likely explanation is that it results from peroxidase activity, as this enzyme has been claimed as a soluble constituent of at least some plant parts.[32] It should be noted that cytoplasmic darkening after incubation in DAB-H_2O_2 medium has been reported for liver cells of several vertebrate species. This phenomenon has been attributed to extraperoxisomal catalase[33,34] and has been offered as evidence for the existence of such activity. So far no cytochemical evidence has been put forth to support the occurrence of extraperoxisomal catalase in plants. This topic should, perhaps, be reinvestigated, particularly in light of current hypotheses that various types of microbodies grow by incorporating enzymes translated on free cytoplasmic ribosomes (see Kunce et al.,[35] for example). Workers should remain aware, however, of earlier reports of catalase-mediated DAB staining on cytoplasmic ribosomes, particularly in animal cells.[36,37] First interpreted as evidence for localization of the newly synthesized enzyme at its translation site, subsequent findings have assigned this staining to a leakage artifact in which catalase diffuses from adjoining peroxisomes and becomes adsorbed nonspecifically to the ribosomes.[38]

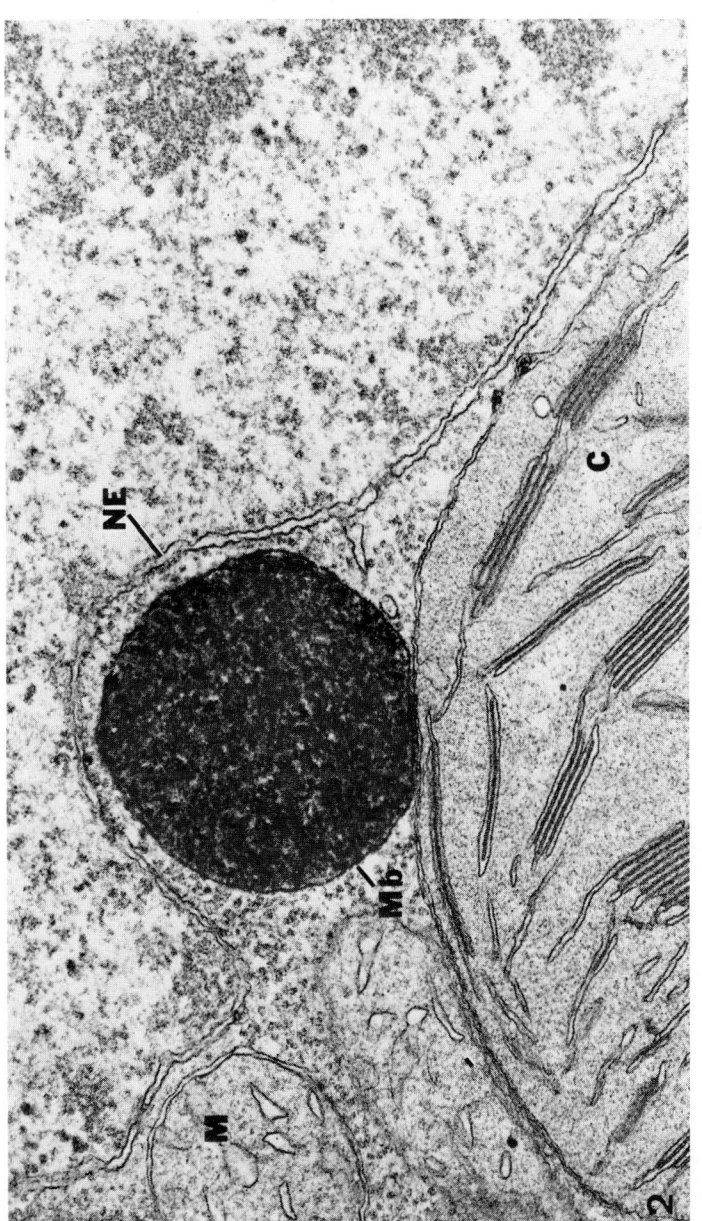

FIGURE 2. Detail of tobacco mesophyll cell incubated in DAB medium designed for visualization of catalase activity. The matrix of the microbody (Mb) shows heavy accumulation of osmium black. This reaction product is lacking in all other parts of the cell, including the nuclear envelope (NE) or nuclear matrix, cytoplasmic matrix, mitochondria (M), and chloroplast (C) shown here. (Magnification × 48,000.) (From Frederick, S. E. and Newcomb, E. H., *J. Cell Biol.*, 43, 343, 1969. With permission.)

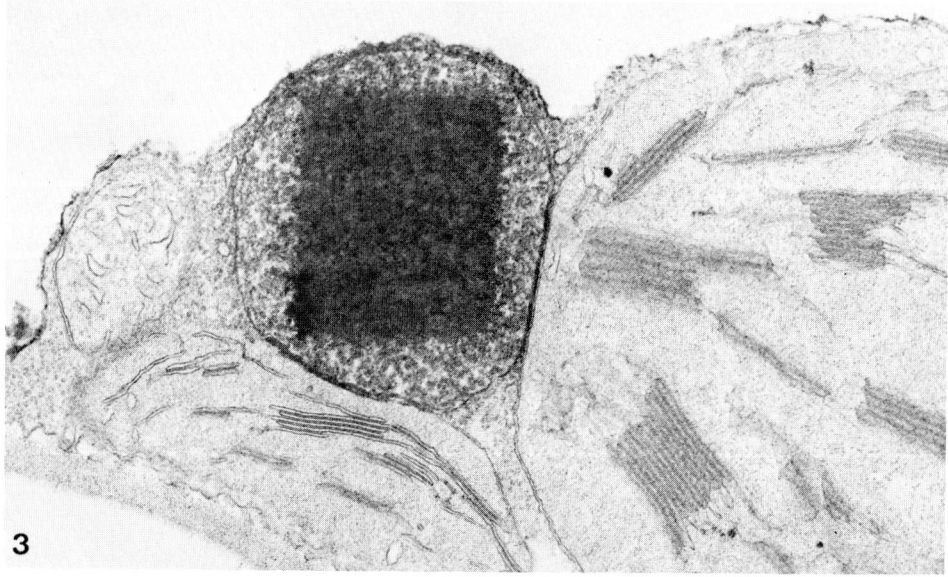

FIGURE 3. Crystal-containing microbody from tobacco leaf mesophyll cell incubated in DAB medium under conditions optimal for revealing catalase activity. The crystal has stained so intensely for catalase activity that reaction product completely obscures the crystal structure. The matrix outside the crystal is much less reactive than the crystal or the matrix of the microbody in Figure 2. (Magnification × 44,000.) (From Frederick, S. E. and Newcomb, E. H., *J. Cell Biol.*, 43, 343, 1969. With permission.)

Identification of Peroxisomes in Algae and Fungi

Applications of DAB cytochemistry to catalase localization in the algae and fungi have been as important as in higher plants for establishing the status of peroxisomes.

DAB staining has now identified microbodies as peroxisomes in species representing all the groups of eukaryotic algae.[39-45] The incubation media and controls used for the localization of catalase have, in general, mimicked those used in higher plants (see Table 1). Results have been much more variable among the algae, however. Attempts to demonstrate catalase in microbody-like organelles in certain species such as *Chlorogonium elongatum* failed in the hands of some investigators,[46] even though they could demonstrate that isolated organelles which were morphologically similar had biochemically detectable catalase activity. Later investigations showed relatively intense catalase staining in microbodies of the same species,[41] the major modification in the technique consisting of an increase in the pH of the medium from 7.4 to 9.4. Apparently, then, the reaction of catalase with DAB in some algae is more sensitive to pH than is that of the enzyme in most higher plants. Cytochemical studies of catalase activity in microbodies of other algae, such as *Nitella flexilis*, have yielded contradictory results even in the hands of the same investigator (compare Silverberg and Sawa[47] with Silverberg[40]). Finally, species such as *Chlamydomonas reinhardi* and *Pediastrum tetras* seem to possess microbody-like organelles which do not stain with DAB under any conditions used to demonstrate catalase in other species,[40,48,49] whereas still others, such as *Porphyridium*, contain microbodies which stain with DAB under conditions favoring catalase reaction, but show inhibition patterns more consistent with peroxidase.[50]

No explanation for some of the DAB staining anomalies can be confirmed at this time. One should be aware of them, however, as they point to possible limitations in the DAB technique as an infallible tag for catalase localization in all species. A more thorough assessment of biochemical variations among the algal catalases should help clarify the situation. Also, spectrophotometric assays of DAB oxidation by cell-free fractions of species in which DAB-negative microbodies occur might help to establish conditions most favorable for DAB staining.

In the fungi as well, DAB has proved useful for identification of microbodies as peroxisomes. As discussed in a recent monograph,[4] microbodies have quite variable morphologies among the fungi as a whole. Hence, the ability to identify various subpopulations of them as catalase-containing peroxisomes has been highly valuable.

The media and inhibitors used so far to localize catalase in fungi are more-or-less identical to those used in higher plants and algae. When staining is present under conditions designed to optimize catalase reactivity (i.e., a high pH and relatively high H_2O_2 concentrations), it seems to be localized exclusively in microbody-type organelles — usually throughout the matrix or concentrated in various inclusion bodies. In general, the matrix staining of fungal microbodies is totally eliminated by AT and to a greater or lesser extent by omission of H_2O_2 from the reaction medium or by the presence of KCN or azide.[51-53]

As in the algae, however, there have existed anomalies in fungal catalase localization.[4,52] Not all structures with microbody morphology have shown DAB staining attributable to catalase (e.g., Mendgen[54]), and microbody-like organelles of some species have exhibited staining and inhibition patterns indicative of the presence of both catalase and peroxidase in the matrix.[52]

Peroxidase

Localizations of endogenous peroxidase activity in plants have, in general, followed the original procedures of Graham and Karnovsky.[1] Typically, aldehyde-fixed material is incubated in a medium of DAB and H_2O_2 buffered to near neutrality. A representative protocol for peroxidase localization is provided in Table 1.

Prefixation

As with catalase, such parameters as prefixation conditions and the pH of the incubation medium are of utmost importance in design of optimal procedures. Unlike catalase, peroxidases actually do not require fixation in order to show reactivity toward DAB.[18] Indeed, some studies of mammalian cells indicate that certain peroxidases are inhibited by aldehyde fixation (by glutaraldehyde more than formaldehyde), and that the milder the aldehyde fixation prior to the cytochemical incubation, the stronger the staining reaction. However, in one of the only plant studies which systematically investigated the effect of fixation, a 3-hr treatment of root tips in relatively strong (3%), cold, buffered glutaraldehyde showed only 11% inhibition of peroxidase activity, measured spectrophotometrically.[55]

Cytochemical Treatment

The optimal pH for peroxidase-mediated DAB oxidation is lower than that for catalase in essentially all reported cases, and media are most commonly buffered around neutrality within the range of 6.5 to 7.5. Depending upon the exact pH chosen, buffers appropriate for the media have included Tris-HCl, cacodylate, and acetate. In spectrophotometric assays of peroxidase activity in pea root homogenates, one group determined the pH optimum to be 5.5 and also obtained good cytochemical localization at this pH.[55] Cytochemical staining of catalase is generally not observed at the lower pH values used for peroxidase detection, although some peroxidases do show DAB oxidation at high pH values and under other conditions used for catalase detection.[18] Thus, while control of pH can favor one or the other enzyme, it alone does not suffice for a rigid distinction between the two.

As for the composition of the DAB reaction mixture, it is H_2O_2 concentration which appears to be of primary importance. Peroxidatic oxidation of DAB by endogenous peroxidases of various mammalian cells seems to be optimal at low H_2O_2 levels (e.g., 0.003%) and to be inhibited by the higher H_2O_2 levels (0.15%) which are most suited for catalase localization.[18,56] This generalization would seem to hold true for plant peroxidases as well.

Incubation times and temperatures vary considerably for peroxidase localization; optimal values depend upon such factors as the extent of penetration required, the prefixatives used prior to the incubation, and so forth. Incubations ranging from 15 to 60 min have been employed at room temperature for tissue pieces,[57] while treatments of 15 min have proved sufficient for thinner Vibratome sections.[55] The degree to which reagents have penetrated should be monitored as described for catalase.

Controls

Controls which can reliably distinguish between cytochemical reactivity of peroxidase and other heme compounds are notably difficult to devise. Unfortunately, peroxidases from different sources, different locations within tissues, or even different regions within single cells seem to vary in their response to inhibitors; this variation may reflect the occurrence of numerous peroxidase isozymes. While the range of controls is quite parallel to that for catalase, i.e., incubation in medium lacking H_2O_2, or addition to the medium of the inhibitors AT, KCN, or azide, results are often more difficult to interpret. The following generalizations are, thus, offered as useful guidelines, not as invariable rules. Omission of H_2O_2 should eliminate peroxidase-mediated staining provided endogenous production of this compound is inhibited (see section on catalase controls). Cyanide, at concentrations of 0.1 or 0.01 M, generally reduces peroxidase staining, but may not eliminate it entirely.[13,57,58] In some cells there seem to be cyanide-sensitive peroxidases (e.g., in the endoplasmic reticulum, Golgi apparatus, cell wall, and cytoplasmic matrix) as well as cyanide-insensitive ones in vacuoles.[13] Aminotriazole (0.02 M) fails to reduce substantially DAB oxidation by most plant peroxidases;[55,57,58] it does, however, inhibit to a greater or lesser extent endogenous peroxidases of animal cells (e.g., see Strum and Karnovsky,[23] Fahimi,[59] and Herzog and Miller[60]).

Applications of Peroxidase Localization

The role of peroxidase(s) in plant cells has remained somewhat clouded. Prerequisite to understanding the function, however, is a knowledge of its distribution within cells and among various cell and tissue types. Though replacement of benzidine with DAB greatly sharpened the resolution of the cytochemical reaction by extending its detection to the electron microscopic level, the literature has retained elements of confusion. Not only is there disagreement among investigators as to the precise distribution pattern, but there seem to be cell, species, tissue, and developmental differences which obscure structure/function generalizations. Furthermore, some problems may arise simply from the inability of the DAB procedure, as currently applied to peroxidase, to resolve fine distinctions between peroxidase and other enzymes which may exhibit similar reactivity toward DAB.

One area in which there remains disagreement is the significance of peroxidase in cell walls. One postulated role for peroxidase is in lignification during secondary wall formation (e.g., see Stafford[61]). One might, thus, expect the enzyme to be localized in areas where lignin is being deposited, and, indeed, it has been reported to be specifically associated with regions of wall thickening during wound vessel formation in *Coleus*, and to a lesser extent with the primary wall and plasma membrane underlying these thickenings.[57] However, with different species others have failed to find peroxidase in lignified xylem, even though unlignified walls in the very same tissues showed peroxidase-mediated DAB oxidation.[62] Certainly, the bulk of evidence would support the presence of peroxidase in primary cell walls which never become lignified, and so if lignification is an important function for peroxidase, it is likely not the only one relating to the cell wall.

Within the cell, DAB staining attributable to peroxidase has been noted in the cytoplasmic matrix by some,[63] but not other,[57] investigators; such a localization would be consistent with cell fractionation techniques which suggest that peroxidase is largely soluble.[32] Failure to

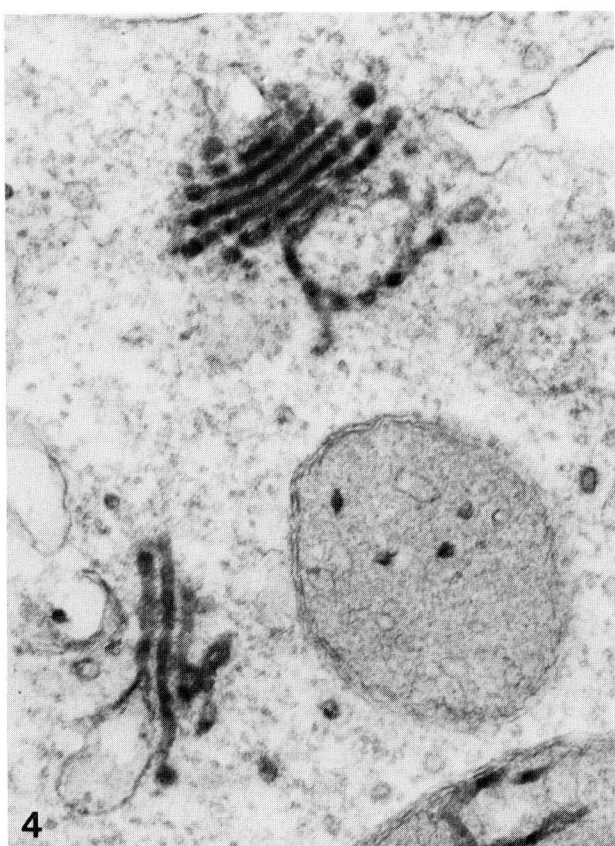

FIGURE 4. Golgi apparatus from protoplast of *Daucus carota* incubated in a DAB medium designed for peroxidase localization. All of the cisternae and many of the associated vesicles are packed with reaction product. In cells such as these, Golgi may be found which exhibit all degrees and many patterns of staining. (Magnification × 69,500.) Micrograph by Sue Ellen Frederick (unpublished).

detect general matrix darkening could reflect a rather low and diffuse concentration of the enzyme rather than its absence.[17] On the other hand, caution must be exercised in ascribing even a low activity of peroxidase to the cytosol, insofar as this and other enzymes may leak from various membrane-bounded particles. Various organelles or cellular compartments have been reported as sites of peroxidase activity — most routinely, the Golgi cisternae and associated vesicles,[57,58,62,64] endoplasmic reticulum,[52,58,62,64] nuclear envelope,[62] ribosomes,[58,63,65] plastid inclusions such as the "thylakoidal body",[66,67] and vacuoles.[13,58,66,68] The localization of peroxidase in the Golgi apparatus, shown in Figure 4, could well reflect the role of this organelle in transporting this enzyme along a typical secretory route to the cell wall. On the other hand, staining of ribosomes seems more likely to be artifactual and due to adsorption of the enzyme which is present in the cytoplasmic matrix.

OTHER ENZYMES OR HEMOPROTEINS DEMONSTRATED BY DAB

Cytochrome Oxidase

Mitochondria of most organisms will, if incubated in an appropriate medium, oxidize DAB and accumulate reaction product on the inner membrane and in the intracristal space

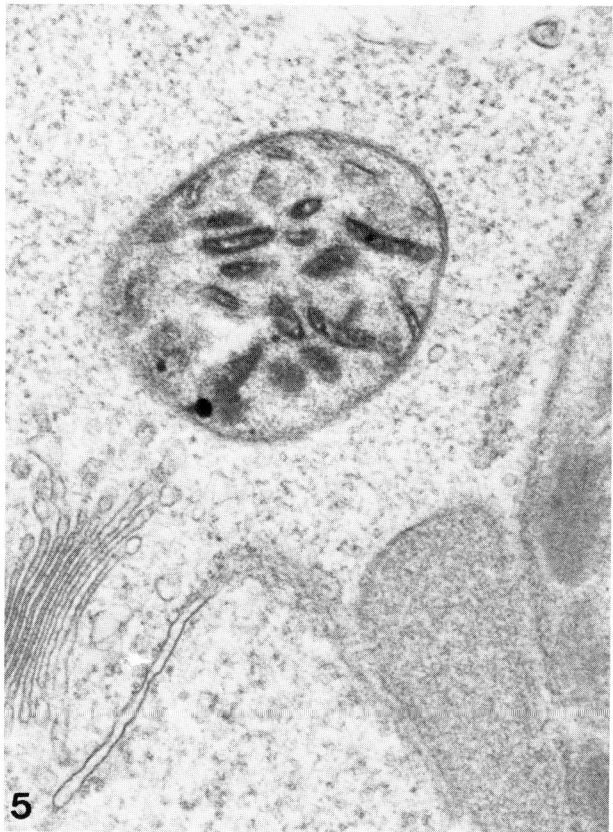

FIGURE 5. Mitochondrion from cell of the green alga *Klebsormidium flaccidum* incubated in a medium containing DAB, but lacking H_2O_2. In this particular case, the reaction product, likely revealing sites of cytochrome oxidase activity, lines the cristae but has not filled the intracristal space. It is not discernible on the noncristal inner membrane either. (Magnification × 55,000.) Micrograph by Nancy Roy (unpublished).

(Figure 5). While it is by no means universally accepted for all organisms and reaction conditions, the original and usual interpretation is that DAB donates electrons to cytochrome *c* and, thence, to the rest of the cytochrome oxidase complex.[7] The reaction, then, both requires *and* shows the presence of cytochrome *c* and, in addition, demonstrates the activity of cytochrome oxidase. Accordingly, procedures which result either in a loss of cytochrome *c* from mitochondria or in an inhibition of cytochrome oxidase activity will reduce the amount of reaction product.

Definitive assurance that mitochondrial DAB staining reveals cytochrome oxidase activity is hard to provide, and often the casual assignment of such staining to this activity cannot be rigorously defended. Indeed, there seem to be several reactions by which mitochondria can oxidize DAB. For example, mitochondria of some organisms, such as yeast, contain a cytochrome *c* peroxidase which apparently is active in oxidizing DAB.[69,70] Also, cytochrome *c* and possibly other cytochromes can act by themselves as peroxidases.[12] And last, some mitochondria such as those in rice coleoptiles[71] oxidize DAB in a manner not due to any known cytochrome system or enzyme activity; while sensitive to cyanide, the reaction is not inhibited by heating to 60°C, a pattern consistent with neither cytochrome nor cytochrome oxidase activity. It should not be surprising, then, that interpretation of mitochondrial DAB staining depends upon careful attention to the details of the procedure, the composition of the medium, and the use of inhibitors.

The magnitude and, indeed, even the precise localization of the reaction are altered by a number of parameters, including the nature and duration of the prefixation prior to the cytochemical incubation. Although aldehyde fixation is generally thought to be damaging to the cytochromes and cytochrome oxidase system and may result in loss of cytochrome c in the case of glutaraldehyde,[7] mitochondria of mammalian tissues fixed in either formaldehyde (4%) or glutaraldehyde (1%)-formaldehyde (4%) for up to 45 min show a DAB reaction product attributable to cytochrome oxidase.[72] The reaction is more intense in unfixed tissues, but here special precautions must be followed to assure adequate morphological preservation. For this purpose short cytochemical incubation times in well-oxygenated medium have been recommended.[72] When fixed tissue is used, long subsequent rinses are prescribed to assure removal of fixing agents which might interfere with the staining reactions.[72] A final point is that the distribution of reaction product on the inner membrane can vary as a function of the fixing conditions.[7,72]

The procedure originally used by Seligman et al.[7] for demonstrating cytochrome oxidase activity in various mammalian organs employed a slightly alkaline medium which contained exogenous cytochrome c (1 mg/mℓ) and catalase (20 µg/mℓ), but lacked H_2O_2. The cytochrome c served to replace that lost from tissue during preparation, while catalase was added to destroy any H_2O_2 which might be generated in the tissue. Few studies have methodically explored the influence of pH on cytochemical reactivity of cytochrome oxidase. While incubations designed specifically to show cytochrome oxidase are usually performed around pH 7.4, the pH optimum for this enzyme, values considerably higher and lower still yield some degree of staining. In one case, the densest deposition of reaction product was found at pH 8.0 to 8.4, but whether this reaction was due solely to cytochrome oxidase activity was not clear.[73] Furthermore, the product was more coarse, more diffuse, and less precisely localized than that at pH 7.2 to 7.4; and the higher pH, it was suspected, might have given rise to some diffusion artifacts of the sort identified by Novikoff et al.[74]

When H_2O_2 is added to the incubation or when no provision is made to remove endogenous H_2O_2, interpretation of mitochondrial staining is rendered more difficult. The possibility then exists that a peroxidase activity is revealed, either of the cytochromes or of specific mitochondrial peroxidases such as cytochrome c peroxidase.[69,70] While, in theory, omission of H_2O_2 should prevent such peroxidase activity, but not that of cytochrome oxidase, it is unwise to rely on this information alone for the distinction between the two.

Incubation times and temperatures optimal for visualizing cytochrome oxidase activity vary, according to the inherent reactivity of the cell type or organism, the size and geometry of the sample (whether tissue blocks, single cells, Vibratome sections, etc.), the conditions of fixation, and the details of the cytochemical medium. The original procedure[7] utilized a 1-hr incubation for fixed mammalian tissue and a 20-min incubation for fresh tissue, but derivative protocols use times as short as 15 min for 1-mm^3 blocks of formaldehyde-fixed tissues, or 30 min for Vibratome sections of formaldehyde/glutaraldehyde-treated tissues. While few detailed cytochemical studies of cytochrome oxidase activity in plants exist, it can be noted that the tendency is toward somewhat longer incubation times, e.g., 1 hr at room temperature for unfixed *Lupinus* root segments[75] and carrot suspension culture cells fixed in cold glutaraldehyde,[63] and usually 1 to 2 hr at 27°C for rice coleoptile segments fixed in glutaraldehyde.[71] It seems likely that these incubation procedures were arbitrarily chosen and do not necessarily reflect the optimal. It would be prudent to consider that these longer incubation times provide more opportunity for diffusion artifacts, and that shorter times would, perhaps, allow a sharper definition of the localization.

Agents which inhibit cytochrome oxidase activity should, in general, serve as controls in determining the mediators of mitochondrial DAB oxidation. Such inhibitors alone are not sufficiently definitive for attributing the reaction to this enzyme, however, as many of them suppress heme protein activity generally. Cytochrome oxidase inhibitors include KCN (usually used at a concentration of about 0.01 M) and azide (used at 3.0 to 6.0 mM), as well as heating to 60°C for 10 min or saturating with carbon monoxide.[72] In some cases, azide only inhibits mitochondrial DAB oxidation after a relatively long preincubation in the presence of this compound.[7]

Photosystem I

DAB can be photooxidized by thylakoid membranes in blue-green algae as well as in chloroplasts of green algae and of higher plants. Patterns of inhibition of this activity lead to the interpretation that it results from donation of electrons by DAB to photosystem I; thus, DAB has emerged as an important marker for this portion of the photosynthetic apparatus. Details of this use of DAB are provided in the chapter on "Photosynthetic Partial Reactions" in this volume.

Polyphenol Oxidase

The usual method for polyphenol oxidase (PPO) localization at the ultrastructural level exploits the ability of this enzyme to oxidize dihydroxyphenylalanine (DOPA) to an osmiophilic o-diquinone (reviewed by Czaninski and Catesson;[76] see, also, Vaughn and Duke[77,78]). There is also some evidence, however, that PPO can utilize DAB as a substrate. Plastid thylakoids are lined with osmium black after incubation with DAB in the dark; the product is eliminated by sodium diethyldithiocarbamate and by KCN, both inhibitors of PPO.[63] Furthermore, the reaction takes place in the absence of H_2O_2 and is not inhibited by agents such as catalase or pyruvate which prevent accumulation of endogenous H_2O_2, or by AT, the catalase suppressor. This inhibition pattern parallels that of PPO-catalyzed oxidation of DOPA and argues against a peroxidase- or catalase-mediated reaction. That the DAB is donating electrons to the photosystems can be ruled out because the reaction occurs in the dark as well as light.

Too few studies have used DAB for demonstrating PPO activity to warrant comparison between it and DOPA as substrates, or to make generalizations regarding favorable reaction conditions. Certainly, the use of DAB would require stringent controls to eliminate the possibility of reactions with peroxidase or the photosystems. The DOPA method is plagued by a similar difficulty, as this compound, too, is acted upon by peroxidase.[76] Regardless of the utility of DAB as a cytochemical probe for PPO, however, it is wise for those interpreting DAB staining of PPO-containing plastids to be aware of DAB's potential reactivity with this enzyme.

DAB CYTOCHEMISTRY AT THE LIGHT MICROSCOPE LEVEL

In nearly all the foregoing citations, DAB has been exploited for localization with the electron microscope, and it has fulfilled this mission well. This section is included as a brief reminder that oxidized DAB and osmium black are discernible in the light microscope as well. In this role, they can be extremely useful in charting the distribution of various heme proteins over relatively large areas. Light microscopic viewing of the DAB reaction product has also led to identification of small, otherwise indistinguishable organelles, as well as to quantitation of enzyme activity via microspectrophotometry.

Identification and Quantitation of Organelles at Low Magnification

DAB staining of microbodies for catalase (or of mitochondria for cytochrome oxidase) allows these organelles to be distinguished from others of similar size and morphology in the light microscope. This capability greatly facilitates quantitation of organelle sizes and volumes, as larger total areas can be scored and analyzed morphometrically than would be possible from electron micrographs at higher magnification. From a single embedment in epoxy resin, one can cut either thin sections for examination in the electron microscope or thick (semithick) sections for surveys under either phase or bright-field optics. An example of the use of such methods is the recent work of Kunce et al.[35] In it, morphometric analysis of numbers and volumes of DAB-stained glyoxysomes at different stages in seed development allowed conclusions to be made regarding the ontogeny of this organelle.

Microspectrophotometry

Investigations by Geerts and Roels[79] revealed that absorbance of oxidized DAB following catalase staining is proportional (in linear fashion) to activity of the enzyme. This observation provided a basis for microspectrophotometric quantitation of such enzyme activity on semithin sections prepared for the light microscope. This method, in combination with morphometric analysis of peroxisome number and size, was recently exploited to determine the proportion of peroxisomal to extraperoxisomal catalase in liver.[80] The information was then applied to questions concerning the site of biosynthesis of this enzyme and the means by which it achieves its ultimate distribution. Such methods have not yet been utilized to explore similar questions in plants.

SOURCES OF ARTIFACT IN DAB CYTOCHEMISTRY

Diffusion

Diffusion of either reaction product or the protein being localized can give rise to a number of artifacts during DAB cytochemical procedures.[74] Some authors doubt that significant migration of oxidized DAB occurs, since over a broad pH range no water-soluble polymer is produced.[81] While the same authors do concede the possibility that small, mobile polymers may be produced at high or even neutral pH, especially by weak oxidizing agents, they are inclined to believe that most artifacts are due to the diffusion of the hemoproteins being localized, particularly when these are not membrane bound. Indeed, it has been well documented that proteins such as catalase can diffuse from their original locations during the buffer rinse prior to cytochemical incubation.[82] For example, prolonged storage in buffer at this stage leads to leakage of catalase from peroxisomes to the adjacent cytoplasm. Such diffusion probably accounts for reports of cytoplasmic ribosomes which stain for catalase activity.[36,37]

Binding of DAB or DAB Oxides to Tissue Components

As unreacted DAB is basic and can potentially interact with acidic components of cells,[81] it represents a possible source of spurious osmium black deposition. Another binding artifact arises from the presence of oxidized DAB in incubation media. DAB is autooxidizable in the presence of UV light and oxygen,[83] and the resulting DAB oxides then seem to attach, apparently with some discrimination, to various hemoproteins such as cytochrome c, catalase, and hemoglobin. This binding can sometimes be inhibited by specific inhibitors of enzyme activity (as is the case with AT, the catalase inhibitor), but it often persists upon prolonged heating which destroys the catalytic activity of the proteins. In order to avoid or minimize this type of artifact one should take care to shield DAB incubation media from light as much as possible, at least in cases where light treatment is not a part of the experimental procedure. As a further precaution, a control may be performed to detect adsorption of DAB or its oxidation products. In the former case, tissues may be prepared as usual and incubated in a saturated solution of DAB under conditions which would inhibit true enzyme activities. Samples are then rinsed, incubated in 3 mM potassium ferricyanide to generate oxidized DAB[1] (which, subsequently, will be observable), rinsed again, postfixed in osmium tetroxide, and processed for electron microscopy.

Inadequate Structural Preservation

The importance of satisfactory preservation to any cytochemical work cannot be overemphasized. Novikoff[84] stresses this point in citing several instances of DAB cytochemistry in which poor preservation has led to faulty or ill-supported conclusions. One must, therefore, pay close attention to the fixation procedure, especially to the selection of aldehyde. Glutaraldehyde is the usual fixative of choice for good structural preservation in cytochemical work. Fortunately, it can be used in most DAB procedures, though, as mentioned earlier, it is destructive to some activities such as cytochrome oxidase. Even in these cases, however, short fixation times usually can be identified which will allow both adequate preservation and some enzyme activity.

FURTHER CONSIDERATIONS

Intensification of DAB Reaction Product

Several strategies have been designed to intensify the reaction product obtained in DAB procedures. The oxidized polymer of DAB is routinely rendered more electron dense by reaction with osmium tetroxide. This procedure has some negative aspects, however, and other reagents have been sought which might create an even greater electron opacity or be easier to apply and handle. The major alternative approaches have included (1) incorporation of other heavy metals such as gold[85] or cobalt plus nickel[86] in place of osmium, and (2) use of a potassium-osmium-cyanide complex (POCC) in place of osmium tetroxide.[87]

Incorporation of either cobalt or gold results in an increased electron density, relative to osmium, in the DAB reaction product. For intensification involving gold, fixed material which has been allowed to oxidize DAB is embedded and sectioned (without osmification); then the grids are immersed in a solution of a salt of the heavy metal (e.g., gold chloride, 1%), rinsed, and air dried.[85] In the case of cobalt incorporation, cobalt chloride and nickel ammonium sulfate are added directly to a DAB incubation medium designed for horseradish peroxidase localization. Neither of these procedures has been attempted on plant materials, to the author's knowledge, and their use in animal research has been minimal. It remains to be seen whether they will prove beneficial in a wide variety of DAB applications.

The POCC procedure[87] involves treatment of DAB-reacted samples with POCC in place of OsO_4. The POCC is allowed to react for 1 hr at room temperature. Localization of DAB reaction product is the same, whether OsO_4 or POCC is used; however, POCC produces a finely granular and more intense product. One drawback to this procedure is that POCC stains smooth membranes intensely, even in the absence of oxidized DAB, and this staining may obscure, or be difficult to distinguish from, the DAB complex.

Form, Quality, and Handling of the DAB Reagent

The purity of commercially available DAB has decreased considerably since its introduction into cytochemistry. This probably owes in part to increased awareness of the health risks involved in its purification and handling. The decrease in purity leads to inconsistency in results, even within the same batch, as small amounts can vary quite widely in composition. One response to this problem has been to weigh out larger amounts of DAB than needed and to store it as a concentrated solution from which small, presumably uniform, aliquots can be drawn.[88] Such solutions, when stored frozen, have provided consistent results for more than a year.

A salt form of DAB, usually the tetrahydrochloride, is preferred to the free base form, as it is much more soluble. This compound, currently most available as a powder, must be handled with utmost care (see below), as it has been found to be a borderline carcinogen.[89] Some companies offer the powder preweighed, in sealed serum bottles, into which can be injected the solvent of choice; this greatly minimizes the possibility of contact or breathing of the powder.

Measures should be taken during all stages of DAB procedures to prevent inhalation or contact with the skin. These precautions may be effected by wearing gloves and handling all the powder and all its solutions in a fume hood. Glassware and instruments which have contacted the powder or solution should be immersed for 24 hr in a strong oxidizing agent such as commercial bleach (sodium hypochlorite) before cleaning by normal procedures.

Alternatives to DAB

The potential carcinogenicity of DAB has led to various searches for alternative cytochemical reagents. Some closely related compounds, N,N'bis (4-amino-phenyl)-1,3-xylylenediamine (BAXD) and N,N'bis (4-amino-phenyl)-N,N'-dimethyl ethylenediamine (BED), are oxidized by tissues to yield osmiophilic polymers.[90,91] The results are different in many respects from those with DAB,[5] however, and these cannot be considered as adequate DAB substitutes.

Substances such as tetramethylbenzidine (TMB)[92] and a combination of p-phenylenediamine and pyrocatechol[93] have been introduced for detection of horseradish peroxidase used as a tracer in animal tissues. So far, these compounds, whose use has recently been extended to peroxidases in plants,[94] have been employed largely, if not exclusively, at the light microscope level. While both have been reported in preliminary studies to yield electron-dense reaction products with peroxidase,[94] the range of their usefulness at the ultrastructural level has yet to be tested. It does not appear that either of these will emerge to be the equal of DAB as regards its generation of osmiophilic reaction products or reactivity with a wide range of hemoproteins.

DAB Cytochemistry in Isolation of Cell Constituents

A recent and novel application of DAB cytochemistry has been in the purification of organelles which contain some DAB-reactive hemoprotein.[95,96] Incubating liver cell fractions in a DAB medium made it possible to alter the density of organelles containing HRP-conjugated ligands previously internalized by the live cells. The resulting increase in density allowed these structures to be separated on a sucrose density gradient from others of an inherently similar density. While this procedure is potentially useful for isolating any cellular entity which can be specifically labeled by a suitable hemoprotein-conjugated tag, it remains to be seen how wide its applications will be, in animals as well as plants. One obvious drawback in the technique, as currently executed, is that only altered and, therefore, often nonfunctional cell components may be isolated.

CONCLUSION

It is now approaching 20 years since DAB made its striking entrance into the arena of enzyme cytochemistry and provided localizations of distinctive clarity with the electron microscope. In its heyday, it stood at the forefront in charting the distribution of peroxisomes among species and tissues. One is tempted to wonder whether, perhaps, the crest of DAB has passed — whether in the future it will ever generate new discoveries or see significant service aside from its ongoing role as a localizer of the marker peroxidase in immuno- and lectin-cytochemistry. A detractor could argue that the pace of its contributions has slackened in recent years or is even nearing exhaustion. A promoter could point to new applications which manage to emerge with the years. My impression is that DAB remains a reagent with considerable potential in the hands of intelligent users. In areas such as peroxisomal biogenesis, the status of peroxisomes in lower eukaryotes, or electron transport assemblies, DAB may yet be harnessed in unpredictable ways to yield significant results.

REFERENCES

1. **Graham, R. C. and Karnovsky, M. J.**, The early stages of absorption of injected horseradish peroxidase in the proximal tubules of mouse kidney: ultrastructural cytochemistry by a new technique, *J. Histochem. Cytochem.*, 14, 291, 1966.
2. **Sexton, R. and Hall, J. L.**, Enzyme cytochemistry, in *Electron Microscopy and Cytochemistry of Plant Cells*, Hall, J. L., Ed., Elsevier/North-Holland, Amsterdam, 1978, chap. 2.
3. **Gahan, P. B.**, *Plant Histochemistry and Cytochemistry. An Introduction*, Academic Press, London, 1984, chap. 5.
4. **Huang, A. H. C., Trelease, R. N., and Moore, T. S.**, *Plant Peroxisomes*, Academic Press, New York, 1983, chap. 2.
5. **Essner, E.**, Hemoproteins, in *Electron Microscopy of Enzymes*, Vol. 2, Hayat, M. A., Ed., Van Nostrand Reinhold, New York, 1974, chap. 1.
6. **Litwin, J. A.**, Histochemistry and cytochemistry of 3,3'-diaminobenzidine. A review, *Folia Histochem. Cytochem.*, 17, 3, 1979.
7. **Seligman, A. M., Karnovsky, M. J., Wasserkrug, H. L., and Hanker, J. S.**, Nondroplet ultrastructural demonstration of cytochrome oxidase activity with a polymerizing osmiophilic reagent, diaminobenzidine (DAB), *J. Cell Biol.*, 38, 1, 1968.
8. **Fahimi, H. D. and Herzog, V.**, A colorimetric method for measurement of the (peroxidase-mediated) oxidation of 3,3'-diaminobenzidine, *J. Histochem. Cytochem.*, 21, 499, 1973.
9. **Hirai, K.-I.**, Specific affinity of oxidized amine dye (radical intermediate) for heme enzymes: study in microscopy and spectrophotometry, *Acta Histochem. Cytochem.*, 1, 43, 1968.
10. **LeHir, M., Herzog, V., and Fahimi, H. D.**, Cytochemical detection of catalase with 3,3'-diaminobenzidine, *Histochemistry*, 64, 51, 1979.
11. **Herzog, V. and Fahimi, H. D.**, The effect of glutaraldehyde on catalase. Biochemical and cytochemical studies with beef liver catalase and rat liver peroxisomes, *J. Cell Biol.*, 60, 303, 1974.
12. **Novikoff, A. B. and Goldfischer, S.**, Visualization of peroxisomes (microbodies) and mitochondria with diaminobenzidine, *J. Histochem. Cytochem.*, 17, 675, 1969.
13. **Poux, N.**, Localisation d'activités enzymatiques dans le méristgème radiculaire de *Cucumis sativus* L. IV. Réactions avec la diaminobenzidine mise en évidence de peroxysomes, *J. Microsc.*, 14, 183, 1972.
14. **Herzog, V. and Fahimi, H. D.**, An improved cytochemical method for demonstration of the peroxidatic activity of beef liver catalase (BLC), *J. Histochem. Cytochem.*, 21, 499, 1973.
15. **Goldfischer, S. and Essner, E.**, Further observations on the peroxidatic activities of microbodies (peroxisomes), *J. Histochem. Cytochem.*, 17, 681, 1969.
16. **Hanker, J. S., Anderson, W. A., and Bloom, F. E.**, Osmiophilic polymer generation catalysis by transition metal compounds in ultrastructural cytochemistry, *Science*, 175, 991, 1972.
17. **Frederick, S. E. and Newcomb, E. H.**, Cytochemical localization of catalase in leaf microbodies (peroxisomes), *J. Cell Biol.*, 43, 343, 1969.
18. **Roels, F., Wisse, E., De Prest, B., and van der Meulen, J.**, Cytochemical discrimination between catalases and peroxidases using diaminobenzidine, *Histochemistry*, 41, 281, 1975.
19. **Bieglmayer, C., Ruis, H., and Graf, J.**, Cytochemical localization of catalase activity in glyoxysomes from castor bean endosperm, *Plant Physiol.*, 53, 276, 1974.
20. **Margoliash, E. and Novogrodsky, A.**, A study of the inhibition of catalase by 3-amino-1:2:4-triazole, *Biochem. J.*, 68, 468, 1958.
21. **Rechcigl, M. and Evans, W. H.**, Role of catalase and peroxidase in metabolism of leucocytes, *Nature*, 199, 1001, 1963.
22. **Longo, G. P., Dragonetti, C., and Longo, C. P.**, Cytochemical localization of catalase in glyoxysomes isolated from maize scutella, *Plant Physiol.*, 50, 463, 1972.
23. **Strum, J. M. and Karnovsky, M. J.**, Cytochemical localization of endogenous peroxidase in thyroid follicular cells, *J. Cell Biol.*, 44, 655, 1970.
24. **Cavallo, T.**, Cytochemical localization of endogenous peroxidase activity in renal medullary collecting tubules and papillary mucosa of the rat, *Lab. Invest.*, 34, 223, 1976.
25. **Burris, R. H. and Little, H. N.**, Oxidases, peroxidases, and catalase, in *Respiratory Enzymes*, Lardy, H., Ed., Burgess Publishing, Minneapolis, 1949, 170.
26. **Fahimi, H. D.**, Cytochemical localization of peroxidatic activity of catalase in rat hepatic microbodies (peroxisomes), *J. Cell Biol.*, 43, 275, 1969.
27. **Vigil, E. L.**, Intracellular localization of catalase (peroxidatic) activity in plant microbodies, *J. Histochem. Cytochem.*, 17, 425, 1969.
28. **Marty, M. F.**, Caractérisation cytochimique infrastructurale de peroxysomes (= "microbodies" *sensu stricto*) chez *Euphorbia characias*, *C. R. Acad. Sci. Paris*, D268, 1388, 1969.
29. **Frederick, S. E. and Newcomb, E. H.**, Ultrastructure and distribution of microbodies in leaves of grasses with and without CO_2-photorespiration, *Planta*, 96, 152, 1971.

30. **Hilliard, J. H., Gracen, V. E., and West, S. H.**, Leaf microbodies (peroxisomes) and catalase localization in plants differing in their photosynthetic carbon pathways, *Planta,* 97, 93, 1971.
31. **Berger, C. and Gerhardt, B.**, Charakterisiering der Microbodies aus Spadix-Appendices von *Arum maculatum* L. und *Sauromatum guttatum* Schott, *Planta,* 96, 326, 1971.
32. **Tolbert, N. E., Kisaki, T., Hageman, R. H., and Yamazaki, R. K.**, Peroxisomes from spinach leaves containing enzymes related to glycolate metabolism, *J. Biol. Chem.,* 243, 5179, 1968.
33. **Roels, F.**, Cytochemical demonstration of extraperoxisomal catalase. I. Sheep liver, *J. Histochem. Cytochem.,* 24, 713, 1976.
34. **Roels, F., de Coster, W., and Goldfischer, S.**, Cytochemical demonstration of extraperoxisomal catalase. II. Liver of rhesus monkey and guinea pig, *J. Histochem. Cytochem.,* 25, 157, 1977.
35. **Kunce, C. M., Trelease, R. N., and Doman, D. C.**, Ontogeny of glyoxysomes in maturing and germinated cotton seeds — a morphometric analysis, *Planta,* 161, 156, 1984.
36. **Legg, P. G. and Wood, R. L.**, New observations on microbodies. A cytochemical study on CPIB-treated rat liver, *J. Cell Biol.,* 45, 118, 1970.
37. **Wood, R. L. and Legg, P. G.**, Peroxidase activity in rat liver microbodies after aminotriazole inhibition, *J. Cell Biol.,* 45, 576, 1970.
38. **Fahimi, H. D.**, Diffusion artifacts in cytochemistry of catalase, *J. Histochem. Cytochem.,* 21, 999, 1973.
39. **Stewart, K. D., Floyd, G. L., Mattox, K. R., and Davis, M. E.**, Cytochemical demonstration of a single peroxisome in a filamentous green alga, *J. Cell Biol.,* 54, 431, 1972.
40. **Silverberg, B. A.**, An ultrastructural and cytochemical characterization of microbodies in the green algae, *Protoplasma,* 83, 269, 1975.
41. **Silverberg, B. A.**, 3,3'-Diaminobenzidine (DAB) ultrastructural cytochemistry of microbodies in *Chlorogonium elongatum*, *Protoplasma,* 85, 373, 1975.
42. **Oliveira, L. and Bisalputra, T.**, Studies in the brown alga *Ectocarpus* in culture: ultrastructural localization of enzymic activities, *Can. J. Bot.,* 54, 913, 1976.
43. **Mesquita, J. F. and Santos, M. F.**, Cytological studies in golden algae (Chrysophyceae). II. First cytochemical demonstration of peroxisomes in Chrysophyceae (*Chrysocapsa epiphytica* Lund), *Cytobiologie,* 14, 38, 1976.
44. **Menzel, D.**, Cytochemischer Nachweis von Katalase in Microbodies bei *Acetabularia mediterranea*, *Planta,* 130, 181, 1976.
45. **Hornung, R. L., Hanzely, L., and Lynch, D. L.**, Occurrence of microbodies in the green alga *Bracteacoccus cinnabarinus* grown heterotrophically, *Protoplasma,* 93, 135, 1977.
46. **Gerhardt, B. and Berger, C.**, Microbodies und Diaminobenzidin-Reaktion in den Acetat-Flagellaten *Polytomella caeca* und *Chlorogonium elongatum*, *Planta,* 100, 155, 1971.
47. **Silverberg, B. A. and Sawa, T.**, An ultrastructural and cytochemical study of microbodies in the genus *Nitella* (Characeae), *Can. J. Bot.,* 51, 2025, 1973.
48. **Giraud, G. and Czaninski, Y.**, Localisation ultrastructurale d'activités oxydasiques chez le *Chlamydomonas reinhardi*, *C. R. Acad. Sci. Paris,* D273, 2500, 1971.
49. **Rogalski, A. A., Overton, J., and Ruddat, M.**, An ultrastructural and cytochemical investigation of the colonial green alga *Pediastrum tetras* during zoospore formation, *Protoplasma,* 91, 93, 1977.
50. **Oakley, B. R. and Dodge, J. D.**, The ultrastructure and cytochemistry of microbodies in *Porphyridium*, *Protoplasma,* 80, 233, 1974.
51. **van Dijken, J. P., Veenhuis, M., Vermeulen, C. A., and Harder, W.**, Cytochemical localization of catalase activity in methanol-grown *Hansenula polymorpha*, *Arch. Microbiol.,* 105, 261, 1975.
52. **Powell, M. J.**, Ultrastructural cytochemistry of the diaminobenzidine reaction in the aquatic fungus *Entophlyctis variabilis*, *Arch. Microbiol.,* 114, 123, 1977.
53. **Hänssler, G., Mühlenbacher, D., and Reisener, H.-J.**, Cytochemical localization of microbodies in *Puccinia graminis* var. *tritici*, *Exp. Mycol.,* 5, 209, 1981.
54. **Mendgen, K.**, Microbodies (glyoxysomes) in infection structures of *Uromyces phaseoli*, *Protoplasma,* 78, 477, 1973.
55. **Hall, J. L. and Sexton, R.**, Cytochemical localization of peroxidase activity in root cells, *Planta,* 108, 103, 1972.
56. **Angermüller, S. and Fahimi, H. D.**, Selective staining of cell organelles in rat liver with 3,3'-diaminobenzidine, *J. Histochem. Cytochem.,* 31, 230, 1983.
57. **Hepler, P. K., Rice, R. M., and Terranova, W. A.**, Cytochemical localization of peroxidase activity in wound vessel members of *Coleus*, *Can. J. Bot.,* 50, 977, 1972.
58. **Goff, C. W.**, A light and electron microscopic study of peroxidase localization in the onion root tip, *Am. J. Bot.,* 62, 280, 1975.
59. **Fahimi, H. D.**, The fine structural localization of endogenous and exogenous peroxidase activity in Kupffer cells of rat liver, *J. Cell Biol.,* 47, 247, 1970.

60. **Herzog, V. and Miller, F.**, The localization of endogenous peroxidase in the lacrimal gland of the rat during postnatal development: electron microscope cytochemical and biochemical studies, *J. Cell Biol.*, 53, 662, 1972.
61. **Stafford, H. A.**, Factors controlling the synthesis of natural and induced lignin in *Phleum* and *Elodea*, *Plant Physiol.*, 40, 844, 1965.
62. **Czaninski, Y. and Catesson, A.-M.**, Localisation ultrastructurale d'activités peroxidasiques dans les tissus conducteurs végétaux au cours du cycle annuel, *J. Microsc.*, 7, 875, 1969.
63. **Olah, A. F. and Mueller, W. C.**, Ultrastructural localization of oxidative and peroxidative activities in a carrot suspension cell culture, *Protoplasma*, 106, 231, 1981.
64. **Poux, N.**, Localisation d'activités enzymatiques dans les cellules du méristèmes radiculaire de *Cucumis sativus* L. II. Activité peroxydasique, *J. Microsc.*, 8, 855, 1969.
65. **Czaninski, Y. and Catesson, A.-M.**, Activités peroxydasiques d'origines diverses dans les cellules d'*Acer pseudoplatanus* (tissus conducteurs et cellules en culture), *J. Microsc.*, 9, 1089, 1970.
66. **Henry, E. W.**, Peroxidase in tobacco abscission zone tissue. III. Ultrastructural localization in thylakoids and membrane-bound bodies of chloroplasts, *J. Ultrastruct. Res.*, 52, 289, 1975.
67. **Hurkman, W. J. and Kennedy, G. S.**, Development and cytochemistry of the thylakoid body in tobacco chloroplasts, *Am. J. Bot.*, 64, 86, 1977.
68. **Griffing, L. R. and Fowke, L. C.**, Intracellular vacuole differentiation in soybean tissue culture cells and protoplasts, *Plant Physiol.*, 72, 16s, 1983.
69. **Todd, M. M. and Vigil, E. L.**, Cytochemical localization of peroxidase activity in *Saccharomyces cerevisiae*, *J. Histochem. Cytochem.*, 20, 344, 1972.
70. **Hirai, K.-I.**, Distribution of peroxidase activity in *Tetrahymena pyriformis* mitochondria, *J. Histochem. Cytochem.*, 22, 189, 1974.
71. **Öpik, H.**, The reaction of mitochondria in the coleoptiles of rice (*Oryza sativa* L.) with diaminobenzidine, *J. Cell Sci.*, 17, 43, 1975.
72. **Anderson, W. A., Bara, G., and Seligman, A. M.**, The ultrastructural localization of cytochrome oxidase via cytochrome c, *J. Histochem. Cytochem.*, 23, 13, 1975.
73. **Hoshino, Y. and Shannon, W. A.**, Mitochondrial oxidation of 3,3'-diaminobenzidine at various pH values, *Acta Histochem. Cytochem.*, 12, 206, 1979.
74. **Novikoff, A. B., Novikoff, P. M., Quintana, N., and Davis, C.**, Diffusion artifacts in 3,3'-diaminobenzidine cytochemistry, *J. Histochem. Cytochem.*, 20, 745, 1972.
75. **Ekés, M.**, The use of diaminobenzidine (DAB) for the histochemical demonstration of cytochrome oxidase activity in unfixed plant tissues, *Histochemie*, 27, 103, 1971.
76. **Czaninski, Y. and Catesson, A.-M.**, Polyphenoloxidases (plants), in *Electron Microscopy of Enzymes*, Vol. 2, Hayat, M. A., Ed., Van Nostrand Reinhold, New York, 1974, chap. 3.
77. **Vaughn, K. C. and Duke, S. O.**, Tissue localization of polyphenol oxidase in *Sorghum*, *Protoplasma*, 108, 319, 1981.
78. **Vaughn, K. C. and Duke, S. O.**, Tentoxin effects on *Sorghum*: the role of polyphenol oxidase, *Protoplasma*, 110, 48, 1982.
79. **Geerts, A. and Roels, F.**, Quantitation of catalase activity by microspectrophotometry after diaminobenzidine staining, *Histochemistry*, 72, 357, 1981.
80. **Geerts, A., De Prest, B., and Roels, F.**, On the topology of the catalase biosynthesis and -degradation in the guinea pig liver. A cytochemical study, *Histochemistry*, 80, 339, 1984.
81. **Seligman, A. M., Shannon, W. A., Hoshino, Y., and Plapinger, R. E.**, Some important principles in 3,3'-diaminobenzidine ultrastructural cytochemistry, *J. Histochem. Cytochem.*, 21, 756, 1973.
82. **Fahimi, H. D.**, Effect of buffer storage on fine structure and catalase cytochemistry of peroxisomes, *J. Cell Biol.*, 63, 675, 1974.
83. **Hirai, K.-I.**, Comparison between 3,3'-diaminobenzidine and auto-oxidized 3,3'-diaminobenzidine in the cytochemical demonstration of oxidative enzymes, *J. Histochem. Cytochem.*, 19, 434, 1971.
84. **Novikoff, A. B.**, DAB cytochemistry: artifact problems in its current uses, *J. Histochem. Cytochem.*, 28, 1036, 1980.
85. **Newman, G. R., Jasani, B., and Williams, E. D.**, Metal compound intensification of the electron-density of diaminobenzidine, *J. Histochem. Cytochem.*, 31, 1430, 1983.
86. **Adams, J. C.**, Heavy metal intensification of DAB-based HRP reaction product, *J. Histochem. Cytochem.*, 29, 775, 1981.
87. **Hoshino, Y., Shannon, W. A., and Seligman, A. M.**, A study of potassium-osmium-cyanide complex application to ultrastructural cytochemistry with diaminobenzidine, *Acta Histochem. Cytochem.*, 10, 172, 1977.
88. **Pelliniemi, L. J., Dym, M., and Karnovsky, M. J.**, Peroxidase histochemistry using diaminobenzidine tetrahydrochloride stored as a frozen solution, *J. Histochem. Cytochem.*, 28, 191, 1980.

89. **Griswold, D. P., Casey, A. E., Weisburger, E. K., and Weisburger, J. H.**, The carcinogenicity of multiple intragastric doses of aromatic and heterocyclic nitro or amino derivatives in young female Sprague-Dawley rats, *Cancer Res.*, 28, 924, 1968.
90. **Seligman, A. M., Wasserkrug, H. L., and Plapinger, R. E.**, Comparison of the ultrastructural demonstration of cytochrome oxidase activity with three bis(phenylenediamines), *Histochemie*, 23, 63, 1970.
91. **Nir, I. and Seligman, A. M.**, Ultrastructural localization of oxidase activities in corn root tip cells with two new osmiophilic reagents coupled to diaminobenzidine, *J. Histchem. Cytochem.*, 19, 611, 1971.
92. **Mesulam, M.-M.**, Tetramethylbenzidine for horseradish peroxidase neurohistochemistry: a non-carcinogenic blue reaction-product with superior sensitivity for visualizing neural afferents and efferents, *J. Histochem. Cytochem.*, 26, 106, 1978.
93. **Hanker, J. S., Yates, P. E., Metz, C. B., and Rustioni, A.**, A new specific, sensitive and non-carcinogenic reagent for the demonstration of horseradish peroxidase, *Histochem. J.*, 9, 789, 1977.
94. **Imberty, A., Goldberg, R., and Catesson, A.-M.**, Tetramethylbenzidine and *p*-phenylenediamine-pyrocatechol for peroxidase histochemistry and biochemistry: two new non-carcinogenic chromogens for investigating lignification process, *Plant Sci. Lett.*, 35, 103, 1984.
95. **Courtoy, P. J., Quintart, J., and Baudhuin, P.**, Shift of equilibrium density induced by 3,3′-diaminobenzidine cytochemistry: a new procedure for the analysis and purification of peroxidase-containing organelles, *J. Cell Biol.*, 98, 870, 1984.
96. **Kay, D. G., Khan, M. N., Posner, B. I., and Bergeron, J. J. M.**, In vivo uptake of insulin into hepatic Golgi fractions: application of the diaminobenzidine-shift protocol, *Biochem. Biophys. Res. Commun.*, 123, 1144, 1984.

CERIUM PRECIPITATION

Albert P. Kausch

Peroxisomes compartmentalize various flavin oxidases which catalyze the transfer of substrate hydrogen to molecular oxygen and, thereby, generate hydrogen peroxide. In most cases, the hydrogen peroxide, subsequently, is degraded in a catalase-mediated reaction, and the majority of cytochemical studies of peroxisomes has relied upon procedures for detection of catalase activity. Until recently, the ferricyanide method[1-4] has been the only procedure for localization of peroxisomal flavin oxidase activity. However, problems with reaction product diffusion and considerable nonspecific deposits limit the usefulness of this technique.

Consequently, Briggs et al.[5] introduced a method employing cerium chloride ($CeCl_3$) to localize an NADH oxidase in human polymorphonuclear leukocytes. Cerium acts as a trapping agent for all endogenous sources of peroxide, precipitating two forms of cerium perhydroxide ($Ce[OH]_2OOH$ and $Ce[OH]_3OOH$) produced by two separate reactions.[6] In the presence of a large excess of substrate for a specific oxidase, cerium perhydroxide occurs cytochemically as a high-resolution, well-localized, electron-dense reaction product with minimal nonspecific deposition. After its introduction, the cerium chloride procedure, subsequently, was modified to demonstrate peroxisomal localization of α-hydroxy acid oxidase, D-amino acid oxidase, and methanol oxidase in *Hansenula polymorpha* yeast cells.[7] The $CeCl_3$ technique has since been successfully used to localize α-hydroxy acid oxidase, D-amino acid oxidase, alcohol oxidase, and urate oxidase in peroxisomes of various animal cell types,[8-12] as well as glycolate oxidase and urate oxidase in peroxisomes of higher plant cells.[13-17]

The purposes of this chapter are to present the cerium chloride method for localization of plant peroxisomal enzyme activity and evaluate its use for conventional transmission electron microscopy (CTEM), scanning, transmission electron microscopy (STEM), scanning electron microscopy with transmitted electron detection (SEM/TED), X-ray microanalysis (XRMA), and high-voltage electron microscopy (HVEM). The generation of substrate-independent deposits and possible artifacts is also discussed.

The cytochemical procedure for cerium localization of peroxisomal enzyme activity, outlined below, essentially is that described by Veenhuis and co-workers[7-9] modified for plant tissues.[13-16]

Aldehyde fixation prior to the cytochemical reaction is desirable for retaining ultrastructural integrity. Cells or small tissue segments (less than 0.5 mm³) are fixed in formaldehyde-glutaraldedyhe (4 to 1% v/v) in 50 mM K-phosphate buffer (pH 6.9) at 4°C. Formaldehyde is used because of its rapid penetration characteristic, and should be prepared fresh from paraformaldehyde immediately before use. Aldehyde fixation should proceed for 5 to 35 min starting when plant materials are first exposed to fixative, followed by three 5-min rinses with 50 mM K-phosphate buffer (pH 6.9) at 21°C and one 20-min wash in 100 mM Tris-maleate (pH 7.5).

Certain peroxisomal enzymes (i.e., urate oxidase) are known to be extremely labile to aldehyde fixation. Therefore, the prefixation step must be omitted or greatly reduced in time.

The aldehyde-fixed or -unfixed samples are then preincubated for 1 hr in a solution containing 100 mM Tris-maleate, 5 mM $CeCl_3 \cdot 7H_2O$, and 50 mM 3-amino-1,2,4 triazole (a potent inhibitor of endogenous catalase activity). The tissues are then incubated at 21°C in the same medium supplemented with 50 mM substrate (i.e., sodium urate, sodium glycolate, methanol, D-alanine, Palmitoyl-CoA, or sodium DL-lactate). To avoid precipitation of reaction components with sodium urate, two separate reaction mixture solutions are made consisting of (1) 6.0 mℓ of 50 mM $CeCl_3 \cdot 7H_2O$ and 50 mM 3-amino-1,2,4 triazole in 100 mM Tris-maleate and (2) 40 mℓ of 5 mg/mℓ sodium urate in 40°C doubled-distilled water. These solutions are mixed dropwise with constant gentle stirring. The pH of preincubation and incubation media should be adjusted to approximate the optima for specific enzyme activities. To avoid carbonate precipitation, double-glass-distilled water used for all solutions must be boiled prior to use and pH adjusted with fresh 5 N NaOH. All solutions containing $CeCl_3$ are filtered (0.45 μm) and thoroughly aerated with CO_2-free air throughout use.

The duration of incubation may be from 1 to 36 hr, depending upon the relative level of enzyme activity in the specific peroxisomes under investigations. For example, glycolate oxidase and urate oxidase activities are similar, and both inherently low in unspecialized peroxisomes of many plant cells[18,19] and, therefore, require extended incubation times for adequate localization. However, the activity of glycolate oxidase in leaf-type peroxisomes may be 100-fold higher than in unspecialized peroxisomes of the same plant,[18] and can be successfully detected with $CeCl_3$ with an incubation time of 2 hr or less. Hourly changes into fresh reaction medium should be made to avoid artifacts that may develop from pH changes or depletion of reaction components.

Control preparations may be made by preincubating samples at 70°C for 4 hr, omission of $CeCl_3$, omission of substrate from reaction mixture, or aldehyde fixation in excess of 18 hr.

Following incubation, specimens are rinsed three times, for 15 min each, in 100 mM sodium cacodylate (pH 6.0) to remove any cerium hydroxide precipitates, and then for 15 min in the same buffer (pH 7.2). Unfixed specimens are then exposed to a formaldehyde-glutaraldehyde (4 to 1% v/v) fixation, same buffer and pH, for 4 hr at 4°C. These tissues are then rinsed three times, 5 min each, with the sodium cacodylate buffer. Postfixation may be either with 1% O_sO_4 (same buffer and pH) or O_sO_4-$K_2Cr_2O_7$ (1 to 2.5%, v/v) for 1 hr and samples are then subdivided for separate TEM and SEM processing.

Specimens for CTEM, SEM/TED, STEM, or HVEM are dehydrated in a graded ethanol-propylene oxide series, embedded in epoxy resin, and sectioned with a diamond knife. Thin sections (silver reflectance) are collected on Formvar-coated slotted copper grids and either viewed with or without poststaining. Poststaining may obscure reaction product in lightly stained organelles. A thin carbon coating over sections on Formvar-coated grids increases their stability under the electron beam. These sections may be viewed with CTEM, SEM/TED, or STEM and the same blocks may be used for HVEM grid preparation. Thick sections, 0.25 to 5.00 μm for HVEM observations, are made from ultrasmooth block faces, collected in slotted copper grids, and transferred to a drop of double-distilled water on a glass slide. The slide is then slightly warmed to expand the sections, which are then transferred to Formvar-coated slotted grids and carbon coated.

The SEM specimens are dehydrated in a graded ethanol series to 100% ethanol, quick frozen in liquid nitrogen, cryofractured, and thawed in 100% ethanol. Samples are then infiltrated with Freon 113 (TF) and critical-point dried with liquid CO_2. Specimens for X-ray microanalysis should be mounted on carbon stubs with carbon (to avoid any X-ray interference) and sputter coated with 150 Å Ag-Pd.

Reaction product from the $CeCl_3$ procedure attributable to glycolate oxidase activity clearly distinguishes leaf-type peroxisomes, exemplified in leaves of *Psychotria* (Figure 1*), as well as unspecialized peroxisomes, in root cortical cells of *Yucca* (Figure 2) from other organelles. Thomas and Trelease[13] have also shown the applicability of this procedure to localization

---

* Key to labeling in the figures: C—chloroplasts; c—calcium oxalate crystal; Ci—crystal idioblast; Cm—chloroplast envelope; CW—cell wall; D—dictyosome; ER—endoplasmic reticulum; G—globule; L—lipid body; M—mitochondria; N—nucleus; P—plastid; S—starch; V—vacuole.

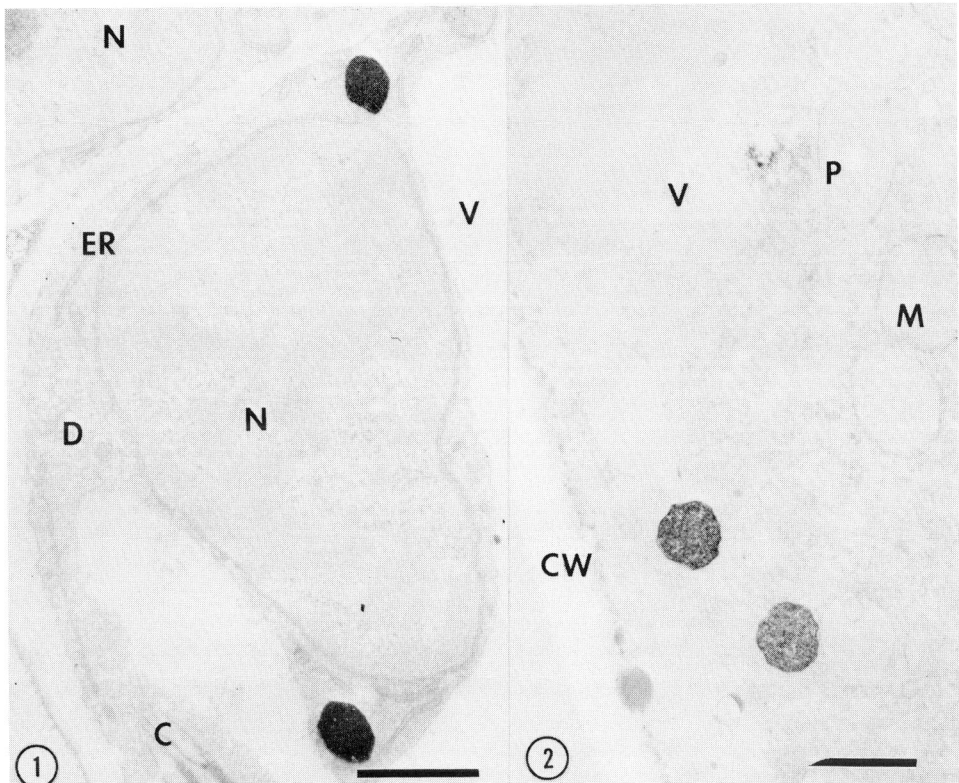

FIGURE 1. Portion of a mesophyll cell from *Psychotria punctata* showing cerium reaction product attributable to glycolate oxidase activity localized to leaf-type peroxisomes. Bar = 1.0 μm. FIGURE 2. Portion of *Yucca torreyi* root cortical cell showing localization of glycolate oxidase activity in unspecialized peroxisomes. Bar = 0.5 μm.

of glyoxysomal glycolate oxidase. Reaction product is throughout the peroxisomal matrix, and nucleoid inclusions become visible as negatively stained structures (Figure 3). The matrix can be observed as contiguous with the interstitial spaces of a crystalline nucleoid inclusions and reaction product occurs in the interstices. These observations demonstrate the high-resolution qualities of the cerium reaction product. Furthermore, these deposits remain well localized and diffusion, or "halo" staining, has not been observed with this procedure. Urate oxidase has been localized with $CeCl_3$ to peroxisomes in root nodules[17] as well as unspecialized peroxisomes (Figures 4 and 5).

Penetrability of reaction components is not complete through entire tissue samples, but limited to a layer, usually 10 to 15 cells inward from the cut tissue surface. It is not clear which compound in the reaction mixture is not well mobilized into plant tissues, however, in studies involving localization of urate oxidase[14] and glycolate oxidase[14-16] penetration was not affected by duration of aldehyde fixation (up to 35 min), preincubation, or reaction medium incubation. Furthermore, increased $CeCl_3$ concentration (up to 50 mM), frequent reaction medium changes, and even pretreatment with 10% dimethoxysulfoxide (DMSO) does not increase penetrability. This is seen as a serious drawback to the cerium procedure, like many other cytochemical procedures, especially concerning studies of peroxisomal distribution and differentiation.[15] All observations involving cerium localization, therefore, must be limited to peripheral cell layers near cut edges of tissue in order to assure reliability.

Variations in staining characteristics among peroxisomes within individual cells have been

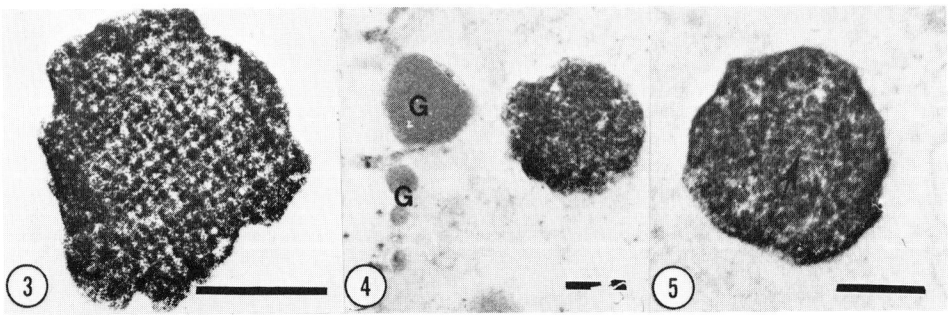

FIGURE 3. High magnification of *Yucca* unspecialized peroxisome reacted for glycolate oxidase activity. Cerium reaction product is throughout peroxisomal matrix, crystalline nucleoid is a negatively stained structure; note the fine, granular high-resolution quality of the strain. Bar = 0.1 μm. FIGURE 4. Positively stained unspecialized peroxisome indicating urate oxidase activity. Globular material accumulates along the plasma membrane in all cells and is likely the consequence of incubation of unfixed tissue. Bar = 0.2 μm. FIGURE 5. Cerium stained unspecialized peroxisome for urate oxidase activity. Smooth appearance of reaction product precipitate may be the result of the high pH of reaction mixture; note the negatively stained crystalline nucleoid inclusion (arrow). Bar = 0.1 μm.

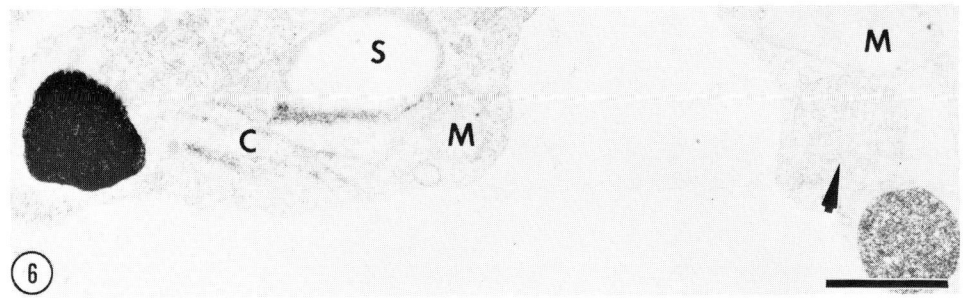

FIGURE 6. Portion of *Glycine* leaf mesophyll cell with a nonreactive peroxisome (arrow) adjacent to a peroxisome containing reaction product and near a very densely stained peroxisome; incubation was 36 hr in 5 mM CeCl$_3$. Bar = 1.0 μm.

frequently observed.[13,14] Figure 6 shows a portion of a *Glycine* cell with a peroxisome that does not contain any reaction product adjacent to a stained peroxisome, and in proximity is a very densely stained peroxisome. This type of variation is observed in cells well within the penetration range of the reaction and is apparent in samples subjected to extended preincubation and/or incubation times, increased CeCl$_3$ concentrations, and DMSO treatment.[14] This variation may indicate the physiological condition of peroxisomes within individual cells reflected by real differences in enzyme activity or content. Such heterogeneity within a cellular peroxisome population would not be easily discerned in biochemical fractionation studies.

Extended incubation times may increase the sensitivity of the procedure, but also introduce various artifacts. Increased incubation times and/or CeCl$_3$ concentrations may result in overloading of reaction product (Figure 7) which makes the material difficult to section. These conditions also result in slight deposition of reaction product in peroxisomes in substrate-minus control samples (Figure 8), indicating the sensitivity of this reaction in the presence of only endogenous substrate.

There are three specific sites of substrate-independent cerium perhydroxide accumulations which have been noted in various plant tissues.[14-16] At incubation times exceeding 4 hr,

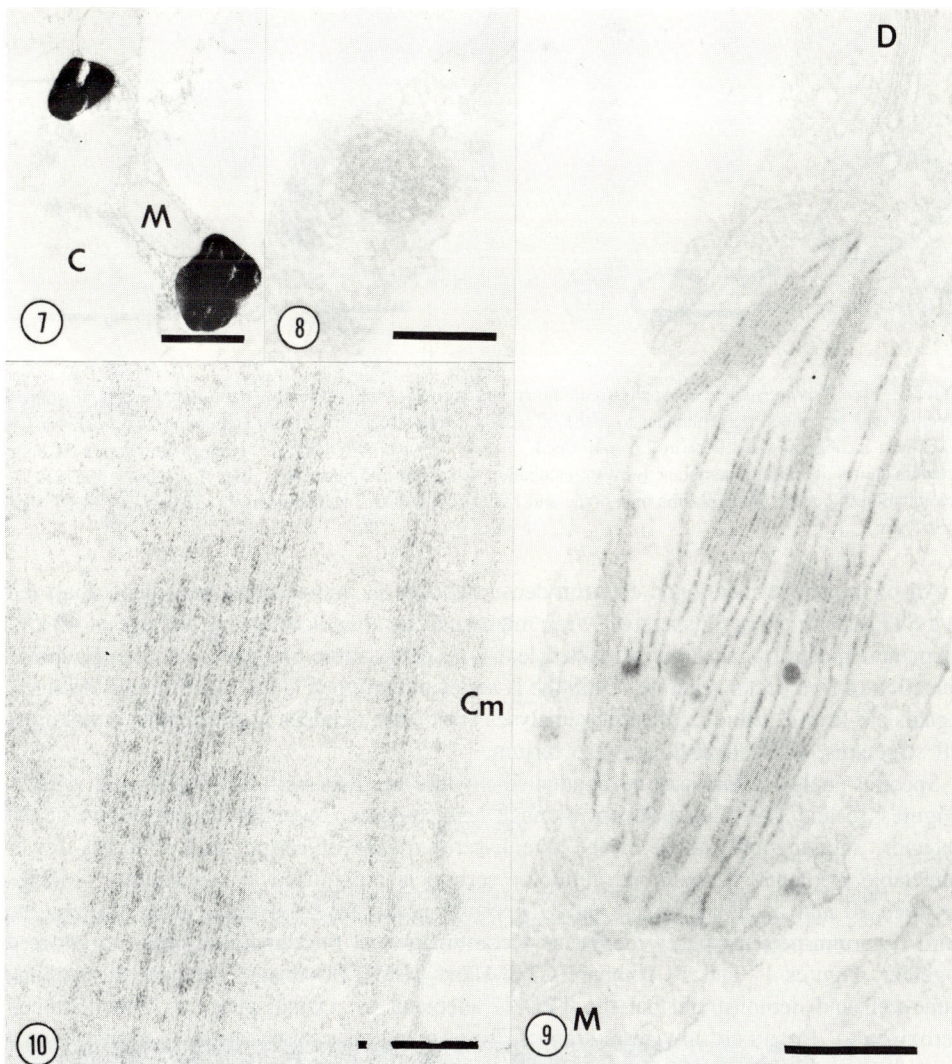

FIGURE 7. Excess deposition of reaction product in *Glycine* resulting in over-stained peroxisomes. Celar areas are where reaction product chipped during sectioning; incubation was 19 hr in 10 mM CeCl$_3$. Bar = 1.0 μm. FIGURE 8. Slight deposition of reaction product in *Psychotria* peroxisome in minus-glycolate substrate control sample, indicating detection of H$_2$O$_2$ in presence of only endogenous glycolate; incubation was 19 hr in 7.5 mM CeCl$_3$. Bar = 0.5 μm. FIGURE 9. Cerium deposition in *Psychotria* chloroplast thylakoids, in minus-glycolate substrate control sample, is confined to thylakoid spaces; incubation was 19 hr in 7.5 mM CeCl$_3$. Bar = 0.1 μm. FIGURE 10. High magnification of *Nicotiana* thylakoid sacs containing cerium deposition; incubation was 11 hr in 5 mM CeCl$_3$. Bar = 0.11 μm.

deposition occurs within thylakoid spaces of chloroplasts (Figures 9 and 10). Deposits are strictly confined to thylakoid spaces of chloroplasts (Figures 9 and 10) and do not accumulate in the stroma. Reduction of oxygen by the chloroplast electron transport system during CO$_2$ assimilation results in the formation of superoxide radicals and/or H$_2$O$_2$ which could produce a cerium perhydroxide pseudo-reaction product. The presence of substrate-independent cerium deposits also has been observed in hoop-shaped structures which resemble ER (Figures 11 and 12) and, also, may indicate the presence of endogenous H$_2$O$_2$. Lastly, some deposits may accumulate within the cell wall (Figure 13) or along the plasma membrane, especially in tissues with minimal intracellular spaces.

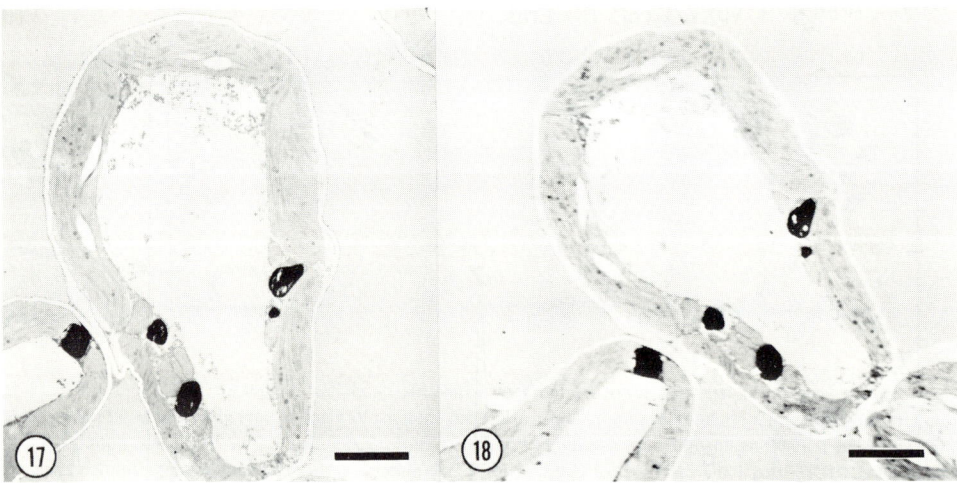

FIGURE 17. Conventional TEM of a thin section through *Glycine* leaf mesophyll cells containing over-stained peroxisomes. Bar = 2.5 μm. FIGURE 18. SEM/TED image of same cells as in Figure 17. Resolution is slightly diminished but cerium deposits still are discerned easily at this magnification. Bar = 2.5 μm.

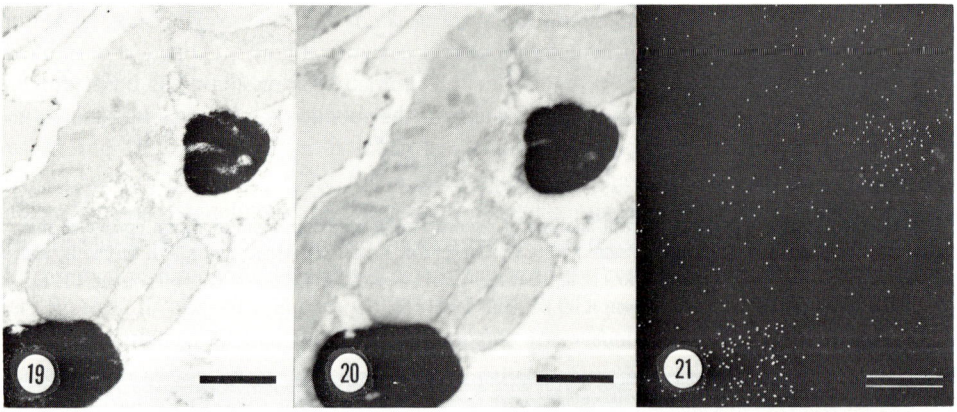

FIGURE 19. High magnification conventional TEM of the same two peroxisomes seen in center cell in Figures 17 and 18. Bar = 0.5 μm. FIGURE 20. SEM/TED image of identical peroxisomes seen in Figure 19. Bar = 0.5 μm. FIGURE 21. SEM/TED X-ray map for the $L_{\alpha 1}$ cerium peak showing peroxisomal localization of reaction product. Bar = 0.5 μm.

Cerium X-ray localizations are accomplished only with significant difficulty, only when beam penetration is minimized using accelerating voltages of 15 to 19 kV and when the cellular structure in question is oriented properly towards the detector. Various tilt angles also are used to minimize background noise from beam penetration. Cerium X-ray localizations depend upon protruding peroxisomes, and thin-sectioned material viewed with CTEM shows that peroxisomes do occasionally protrude into the central vacuole (Figure 28). This may represent an artifact of inadequate fixation, because the tonoplast of this cell is disrupted.

This use of X-ray mapping of cerium reaction products and SEM visualization of plant peroxisomes, therefore, may have only specialized applications. However, this information indicates that cerium-loaded peroxisomes might be easily visualized in the backscattered electron mode. Such visualizations of stained peroxisomes occluded from view by cytoplasm

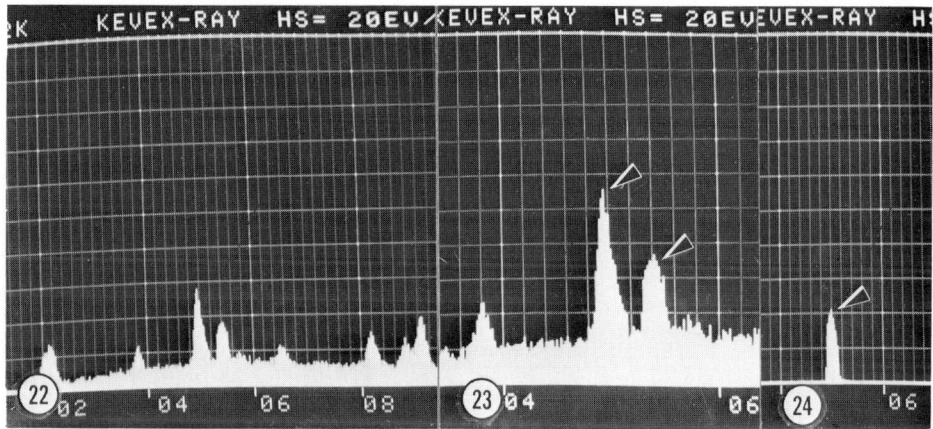

FIGURE 22. Elemental X-ray spectral analysis of cryofractured bulk *Glycine* leaf specimen, analyzed at 20 kV with SEM. FIGURE 23. Enlarged X-ray spectrum shows $L_{\alpha 1}$ and $L_{\beta 1}$ cerium peaks are easily discerned (arrows), whereas remaining L-series peaks are lost in background noise. FIGURE 24. Limitation of $L_{\alpha 1}$ cerium peak for X-ray mapping.

FIGURE 25. Secondary electron image of spherical structure protruding into the vaculoe of a *Glycine* leaf mesophyll cell. Bar = 5.0 μm. FIGURE 26. Higher magnification of structure in Figure 25. Bar = 1.0 μm. FIGURE 27. X-ray map for $L_{\alpha 1}$-cerium signal shows that structure shown in Figures 25 and 26 contains significant amounts of cerium. Bar = 1.0 μm.

or other organelles in SEM-prepared bulk specimens would be ideal for many investigative purposes involving peroxisomal distribution.

Alternatively, up to 5.0 μm of cell depth can be viewed at once in epoxy-embedded thick sections observed with HVEM. Cerium reaction product attributable to enzyme activity is easily discerned in the HVEM operated between 80 and 1000 kV (Figure 29). The HVEM has been shown to be useful for observation of cell-specific localization of peroxisomal glycolate oxidase activity[15] (Figure 30), as well as an investigation of peroxisome biogenesis.[16]

The above discussion shows a wide range of microscopic application for cytochemical procedures involving the $CeCl_3$ technique, based upon the inherent properties of the reaction product. Presently, only two plant peroxisomal oxidases, glycolate oxidase and urate oxidase,

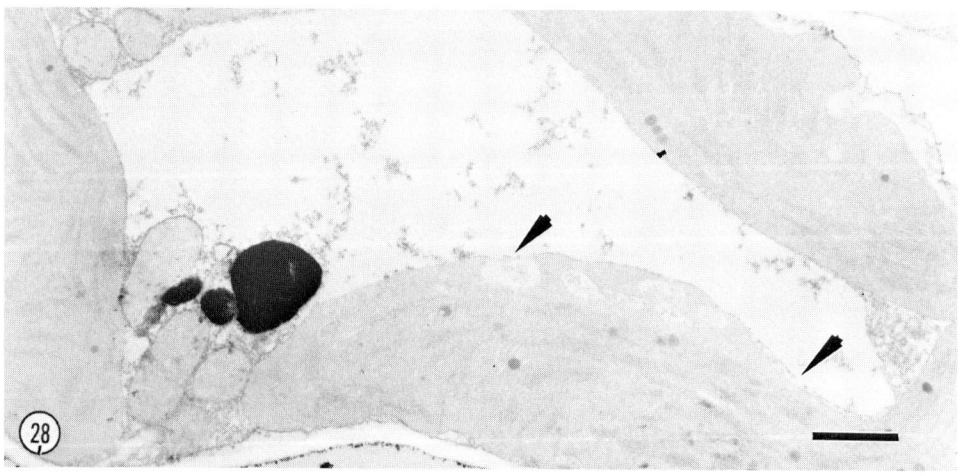

FIGURE 28. Section through reacted *Glycine* leaf mesophyll cell, showing reacted peroxisomes protruding into central vacuole. Note discontinuities in tonoplast (arrows). Bar = 1.0 μm.

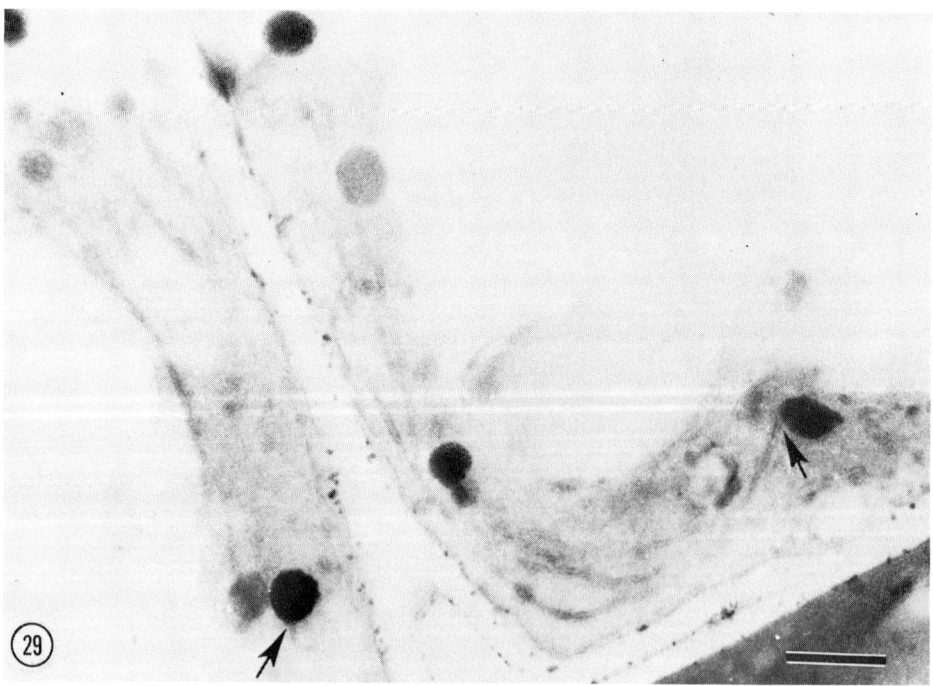

FIGURE 29. Cerium deposition from glycolate oxidose activity in unspecialized peroxisomes of *Yucca* root cortical cell is clearly visible in 2.5-μm thick sections viewed with HVEM. Deposits in wall are not a consequence of enzyme activity. Note apparent ER connections to peroxisomes (arrows). Bar = 1.0 μm.

have been successfully localized using the $CeCl_3$ procedure. Both are easily localized in peroxisomes which contain a high activity of these enzymes. Even though the procedure was first successfully applied in plant cytochemistry of glycolate oxidase,[13] Vaughn and co-workers[17] demonstrated that assessment of fixative effects and reaction conditions for optimal enzyme activity would allow the use of the $CeCl_3$ procedure for cytochemistry of other enzymes. Further studies are necessary to extend the utility of this procedure for identification

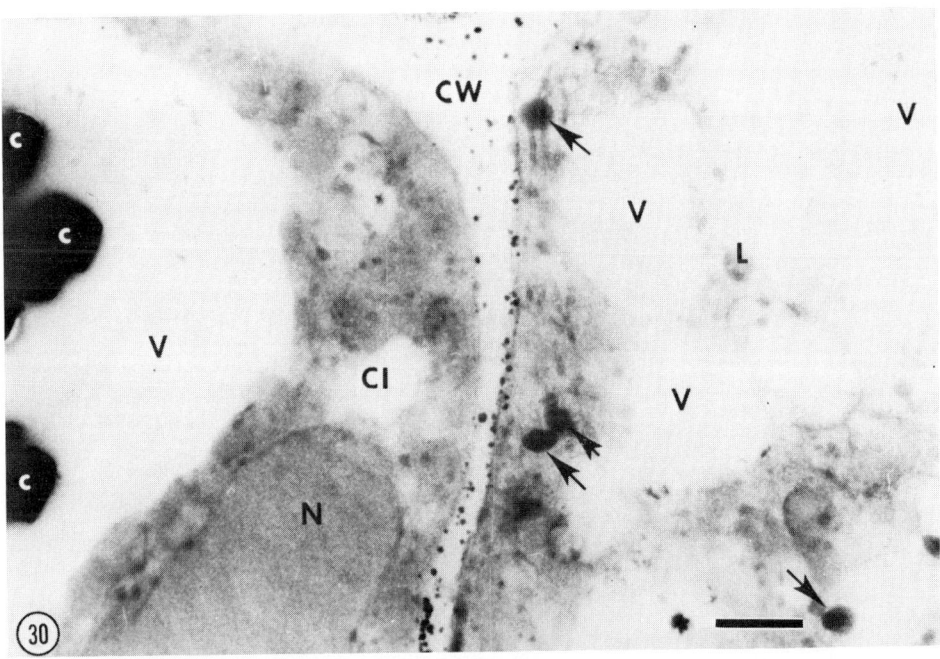

FIGURE 30. Thick section, 5.0 μm, showing absence of cerium stained peroxisomes in crystal idioblast cytoplasm; cerium stained peroxisomes in adjacent cells (arrows) are clearly visible. Bar = 1.0 μm.

of other peroxisomal oxidases, and most certainly must include modifications of the present procedure according to biochemical assessments for optimal reaction conditions.

REFERENCES

1. **Shnitka, T. K. and Talibi, G. G.**, Cytochemical localization by ferricyanide reduction of α-hydroxy acid oxidase activity in peroxisomes of rat kidney, *Histochemie*, 27, 137, 1971.
2. **Burke, J. J. and Trelease, R. N.**, Cytochemical demonstration of malate synthase and glycolate oxidase in microbodies of cucumber cotyledons, *Plant Physiol.*, 56, 710, 1975.
3. **Hand, A. R.**, Ultrastructural localization of L-α-hydroxy acid oxidase in rat liver peroxisomes, *Histochemistry*, 41, 195, 1975.
4. **Hand, A. R.**, Ultrastructural localization of catalase and L-α-hydroxy acid oxidase in microperoxisomes of *Hydra*, *J. Histochem. Cytochem.*, 24, 915, 1976.
5. **Briggs, R. T., Drath, D. B., Karnovsky, M. L., and Karnovsky, M. J.**, Localization of NADH oxidase on the surface of human polymorphonuclear leukocytes by a new cytochemical method, *J. Cell Biol.*, 67, 566, 1975.
6. **Huang, A. H. C., Trelease, R. N., and Moore, T. S., Jr.**, *Plant Peroxisomes*, Academic Press, New York, 1983.
7. **Veenhuis, M., van Dijken, J. P., and Harder, W.**, Cytochemical studies on the localization of methanol oxidase and other oxidases in peroxisomes of methanol-grown *Hansenula polymorpha*, *Arch. Microbiol.*, 3, 123, 1976.
8. **Veenhuis, M. and Wendelaar Bonga, S. E.**, The cytochemical demonstration of catalase and D-amino acid oxidase in the microbodies of teleost kidney cells, *Histochem. J.*, 9, 171, 1977.
9. **Veenhuis, M. and Wendelaar Bonga, S. E.**, The cytochemical localization of catalase and several hydrogen peroxide-producing oxidases in the nucleoids and matrix of rat liver peroxisomes, *Histochem. J.*, 2, 561, 1979.
10. **Arnold, G., Liscum, L., and Holtzman, E.**, Cytochemistry of D-amino acid oxidase in rat cerebellum and kidney, *J. Cell Biol.*, 75, 202A, 1977.

11. **Arnold, G., Liscum, L., and Holtzman, E.,** Ultrastructural localization of D-amino acid oxidase in microperoxisomes of the rat nervous system, *J. Histochem. Cytochem.*, 27, 735, 1979.
12. **Arnold, G. and Holtzman, E.,** Ultrastructural localization of α-OH acid oxidase in peroxisomes with the $CeCl_3$ technique, *J. Histochem. Cytochem.*, 28, 1025, 1980.
13. **Thomas, J. and Trelease, R. N.,** Cytochemical localization of glycolate oxidase in microbodies (glyoxysomes and peroxisomes) of higher plant tissues with the $CeCl_3$ technique, *Protoplasma*, 108, 39, 1981.
14. **Kausch, A. P., Wagner, B. L., and Horner, H. T.,** Use of the cerium chloride technique and energy dispersive X-ray microanalysis in plant peroxisome identification, *Protoplasma*, 118, 1, 1983.
15. **Kausch, A. P. and Horner, H. T.,** Absence of $CeCl_3$-detectable peroxisomal glycolate oxidase activity in developing raphide crystal idioblasts in leaves of *Psychotria punctata* and roots of *Yucca tarreyi*, *Planta*, in press.
16. **Kausch, A. P.,** Biogenesis and cytochemistry of unspecialized peroxisomes in root cortical cells of *Yucca torreyi* L., *Eur. J. Cell Biol.*, 34, 239, 1984.
17. **Vaughn, K. C., Duke, S. O., and Henson, C. A.,** Ultrastructural localization of urate oxidase in nodules of *Sesbania exalta, Glycine max,* and *Medicago satira, Histochemistry*, 74, 309, 1982.
18. **Huang, A. H. C. and Beevers, H.,** Isolation of microbodies from plant tissues, *Plant Physiol.*, 48, 637, 1971.
19. **Huang, A. H. C.,** Metabolism in plant peroxisomes, in *Recent Advances in Phytochemistry*, Creasy, L. L. and Hrazdina, G., Eds., Plenum Press, New York, 1982.

PHOSPHATASES

Randy Moore, C. Edward McClelen, and Houston S. Smith

INTRODUCTION

Phosphatases are a large group of nonspecific enzymes that can hydrolyze phosphate esters with the concomitant release of inorganic phosphate as a reaction product. Phosphatases were first localized cytochemically over 40 years ago,[1,2] and today remain some of the most intensively studied enzymes by cytochemists. There are two general methods for cytochemically localizing phosphatases: (1) capture of the "R" group to which the phosphate was originally attached (e.g., the azo dye method), and (2) capture of the liberated phosphate group with lead ions (Figure 1).

The Azo Dye Method

The azo dye method for cytochemically localizing phosphatases relies on the broad substrate specificity of phosphatases, which includes naphthyl phosphates. Naphthol, a product of this hydrolysis, can combine with a diazonium salt to yield an insoluble, colored precipitate (i.e., the azo dye) at the site of enzymatic activity (Figure 2).

Some investigators have reported that the azo dye method for phosphatase localization is not easily modified for electron microscopic cytochemistry, since the azo dye (1) often obscures ultrastructural detail, (2) is partially soluble in postfixation treatments, and (3) is not very electron-opaque.[3] Furthermore, diazonium salts themselves can inhibit phosphatases in plants.[4] However, there have recently been several improvements in the technique, most notably the development of substituted naphthol phosphate substrates (e.g., naphthol AS-BI phosphate) that allow for very precise localizations. Indeed, Charvat and Esau,[5] using a modified azo dye technique of Bowen,[6] reported ultrastructural localizations of acid phosphatase that (1) were comparable to those resulting from lead-based techniques, and (2) had the advantage of a finer precipitate than lead-based techniques.

Lead-Based Techniques

Lead-based techniques are the most widely used means for cytochemically localizing phosphatases at the ultrastructural level. According to these methods, fixed tissues are incubated in a medium containing a phosphorylated substrate and lead ions, the latter of which immediately trap the liberated phosphate at the site of enzymatic activity (Figure 3). This technique was developed independently by Gomori[1] and Takamatsu,[2] and today is known by the former scientist's name.

Early light microscopists typically converted the insoluble (and difficult to see) lead phosphate reaction product into a black lead sulfide precipitate by treating the tissue with ammonium, sodium, or hydrogen sulfide. Although lead sulfide is a suitable stain for electron microscopy, investigators soon discovered that lead phosphate is also electron dense and easily detectable with the electron microscope. Thus, the conversion of lead phosphate to lead sulfide was deemed unnecessary and has been eliminated. Examples of lead phosphate precipitate resulting from phosphatase localization are shown in Figure 4 and 5.

Although widely used by cytochemists, the validity of the Gomori reaction (Figure 3) for the cytochemical localization of phosphatases has been questioned. The major questions involve (1) the inconsistent staining reaction of the nucleus[7,8] (compare nuclear staining in Figures 5 to 7) and (2) diffusion of the reaction product from its site of production.[9,10] While disagreement regarding the validity of the technique continues,[11] use of the Gomori-type localization for phosphatases has, nevertheless, produced a considerable amount of valuable information about the distribution of phosphatases in plant cells.[8]

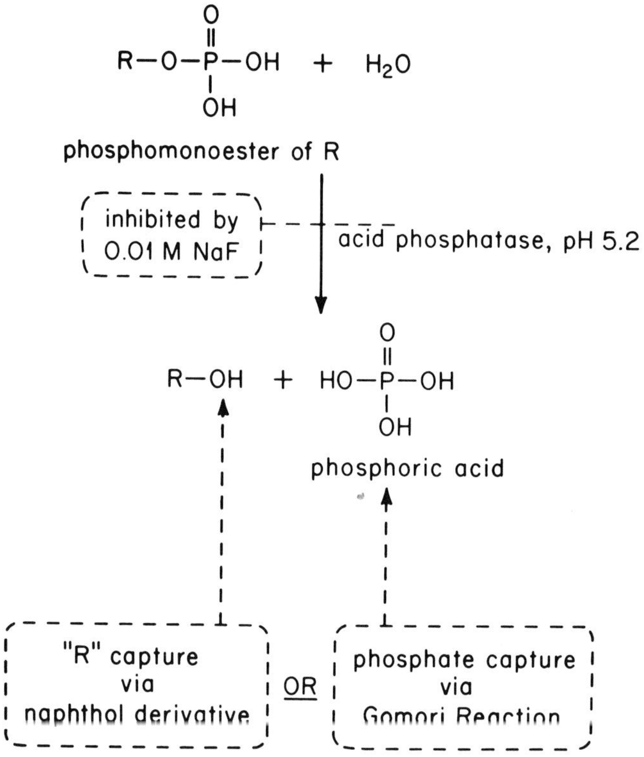

$$R-O-\overset{\overset{O}{\|}}{\underset{\underset{OH}{|}}{P}}-OH \;+\; H_2O$$

phosphomonoester of R

⌐ ‾ ‾ ‾ ‾ ‾ ¬
¦ inhibited by ¦ ── ─ acid¯phosphatase, pH 5.2
¦ 0.01 M NaF ¦
└ ─ ─ ─ ─ ─ ┘
                ↓

$$R-OH \;+\; HO-\overset{\overset{O}{\|}}{\underset{\underset{OH}{|}}{P}}-OH$$

phosphoric acid

┌─────────────┐   ┌─────────────┐
¦ "R" capture ¦   ¦ phosphate capture ¦
¦   via   ¦ OR ¦   via   ¦
¦ naphthol derivative ¦ ¦ Gomori Reaction ¦
└─────────────┘   └─────────────┘

FIGURE 1. The two general methods for cytochemically localizing phosphatases.

FIXATION OF TISSUE

Space limitations do not permit detailed analyses of all factors affecting cytochemical localization of phosphatases. Only those items unique to phosphatases will be discussed here. Readers are referred to other chapters of this book for more generalized discussions of the effects of fixation, postfixation procedures, etc. on enzyme cytochemistry.

Choice of Fixative

Cytochemical localization of enzymes begins with fixation of the tissues to be examined. Chemical fixation of plant tissues serves to immobilize phosphatases in their in vivo positions, thereby reducing their diffusion to other parts of the cell.[12] Aldehydes are usually the preferred fixatives for phosphatase cytochemistry, since osmium and permanganate bring about oxidative cleavage of proteins, thereby completely inactivating the enzymes.[8] However, even aldehydes typically inactivate phosphatases to some degree, depending on which aldehyde is used. Fixation with formaldehyde usually results in less inactivation of phosphatases than glutaraldehyde fixation (Table 1). The extent of inactivation varies with different phosphatases (Table 1). Furthermore, the same phosphatase may have a differential sensitivity to fixation, depending on its cellular location.[13] Interestingly, there are also reports that fixation may activate phosphatases in cultured plant cells.[14]

Phosphatase activity remaining after fixation usually represents only a small fraction of the original activity. The localization of this residual activity is assumed to accurately reflect the distribution of all of the enzyme present prior to fixation.[9]

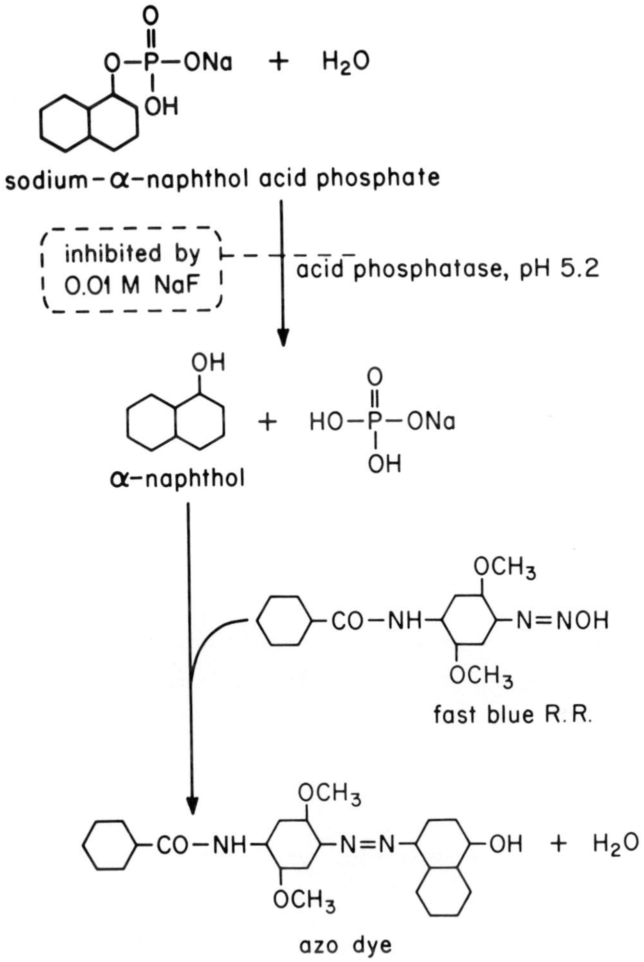

FIGURE 2. The azo dye method for localizing phosphatases.

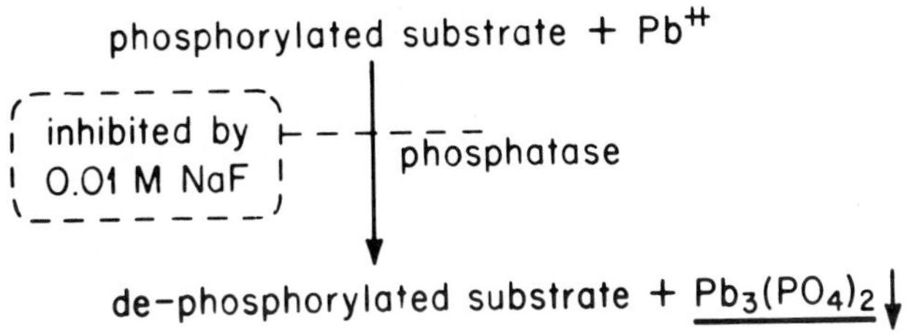

FIGURE 3. The Gomori reaction, a lead-based method for localizing phosphatases.

Duration of Fixation

Maximal inactivation of phosphatases by fixatives typically occurs after 1 to 2 hr. Longer fixation times usually have little effect on phosphatase activity. One exception to this generalization is β-glycerophosphatase, the activity of which gradually disappears during extended fixations (Figure 8).[8]

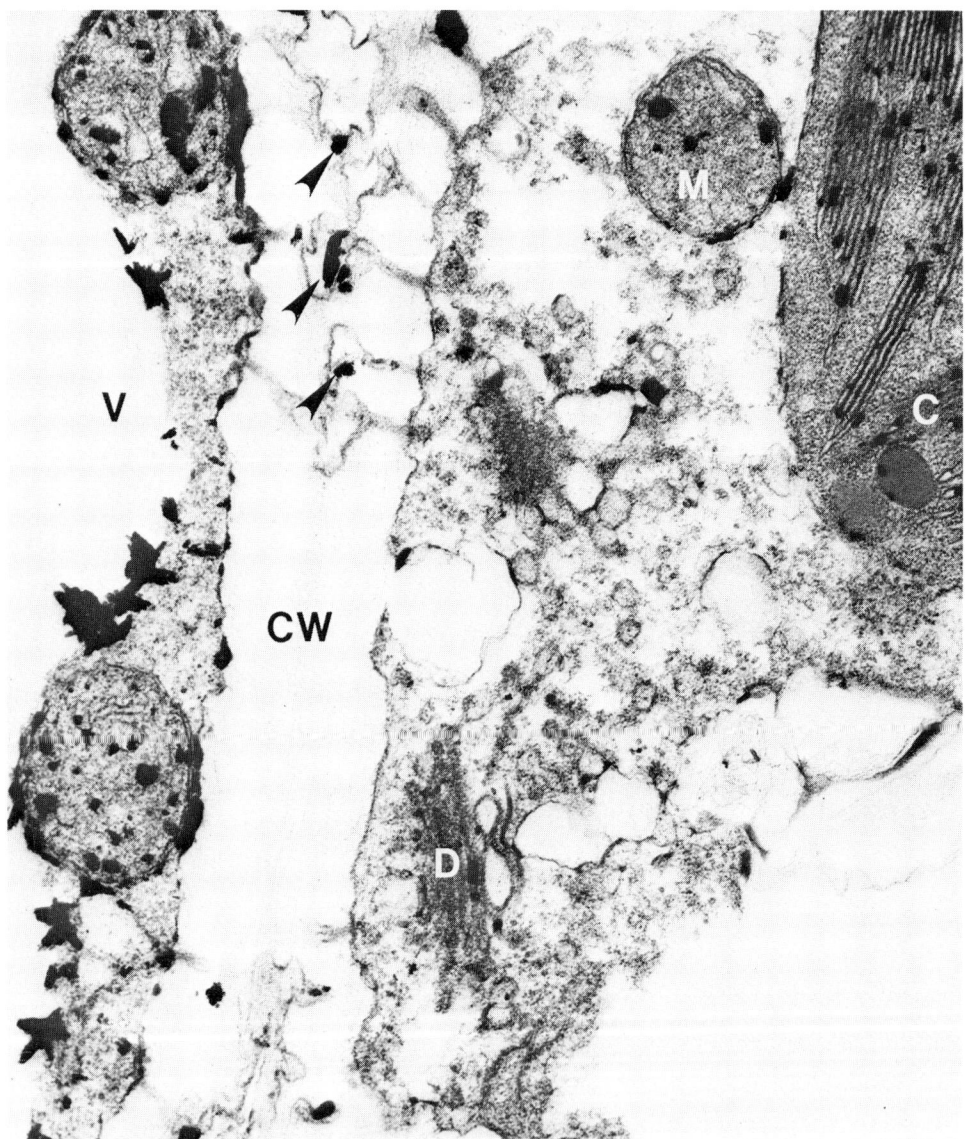

FIGURE 4. Cytochemical staining for acid phosphatase in a stem cortical cell of *Solanum pennellii*. Arrows indicate reaction product in plasmodesmata. C = chloroplast; D = dictyosome; CW = cell wall; M = mitochondria; V = vacuole. (Magnification × 46,500.) (From Moore, R. and Walker, D. B., *Protoplasma*, 109, 317, 1981. With permission.)

Concentration of Fixative

The optimal concentrations of glutaraldehyde and formaldehyde for phosphatase cytochemistry range from 1 to 4%.[8] However, these concentrations are apparently not critical. For example, Goff and Klohs[15] reported that concentrations of glutaraldehyde ranging from 0.5 to 4% had similar effects on the inactivation of nucleoside phosphatase in *Allium* root tips. The chief effect of increasing the concentration of fixative is to decrease the time required to reach maximal inhibition.[15] The rate of phosphatase deactivation during fixation may be decreased by including the substrate in the fixation medium.[16]

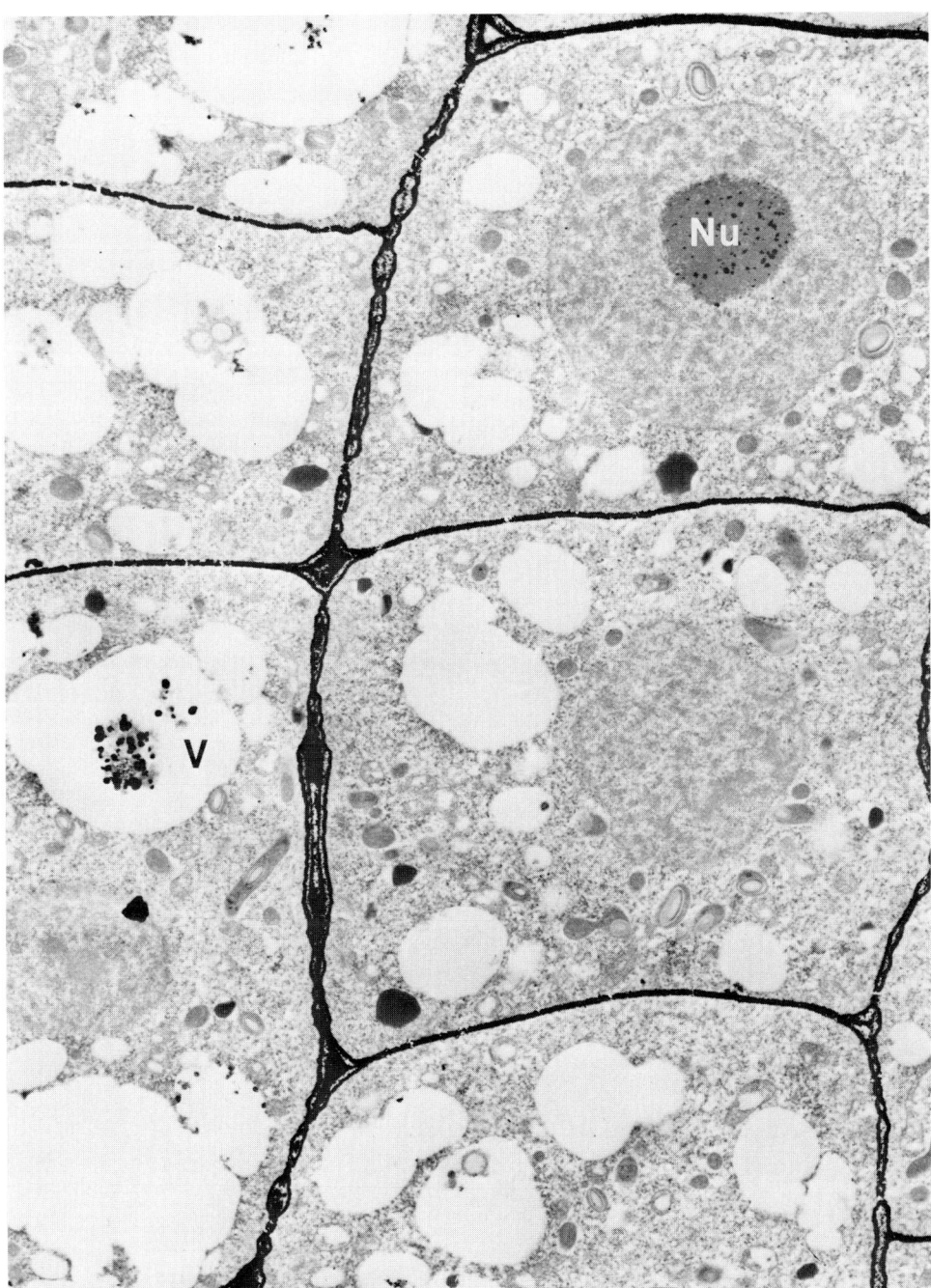

FIGURE 5. Cytochemical staining for acid phosphatase in a root cortical cell of *Zea mays*. Reaction product occurs in the cell wall, plasmalemma, nucleolus (Nu), and vacuole (V). (Magnification × 9800.)

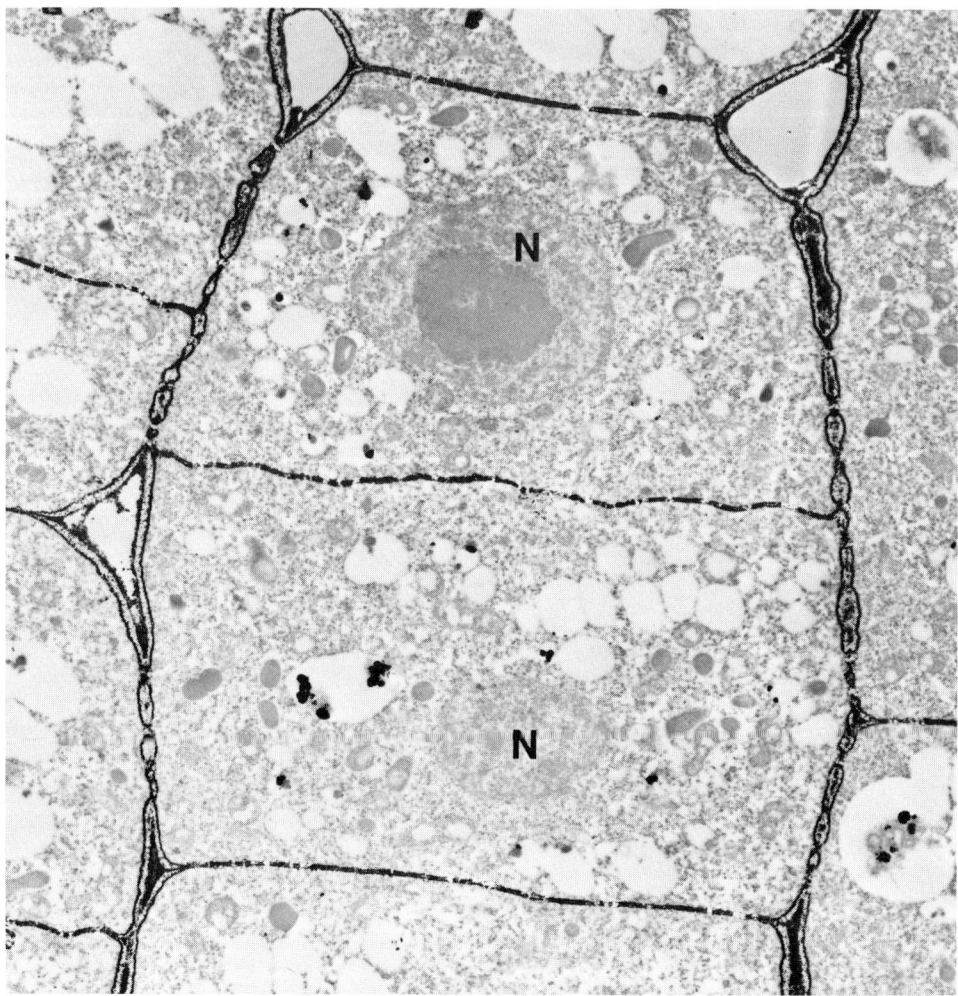

FIGURE 6. Cytochemical staining for acid phosphatase in a root cortical cell of *Zea mays*. Reaction product is located along the plasmalemma and in the cell wall and vacuole. Note the absence of staining in nuclei (N). (Magnification × 9900.)

PENETRATION OF INCUBATION MEDIUM

In order to insure a high-quality localization it is critical that the incubation medium adequately penetrate the tissue. This problem is usually overcome by using thin sections of tissue (i.e., 50 to 70 μm thick). Interestingly, extending the incubation period has little effect on penetration of the incubation medium (Table 2).[8] Rather, increased incubation times result in significant differences in the localization and intensity of cytochemical staining.[8,17,18] An example of a penetration problem associated with the cytochemical localization of thiamine pyrophosphatase (TPPase) is shown in Figure 9. Note that the reaction product is most intense at the cut surface, yet does not penetrate the entire cell. The addition of 0.01% Triton® WR 1339 to the incubation medium has been suggested to allow the cytochemical revelation of enzyme sites not discernible in the absence of Triton®.[19,20]

CONTROLS

Controls are essential for demonstrating that any precipitate is due to the activity of a specific enzyme rather than nonenzymatic or enzymatically produced "false" precipitation.

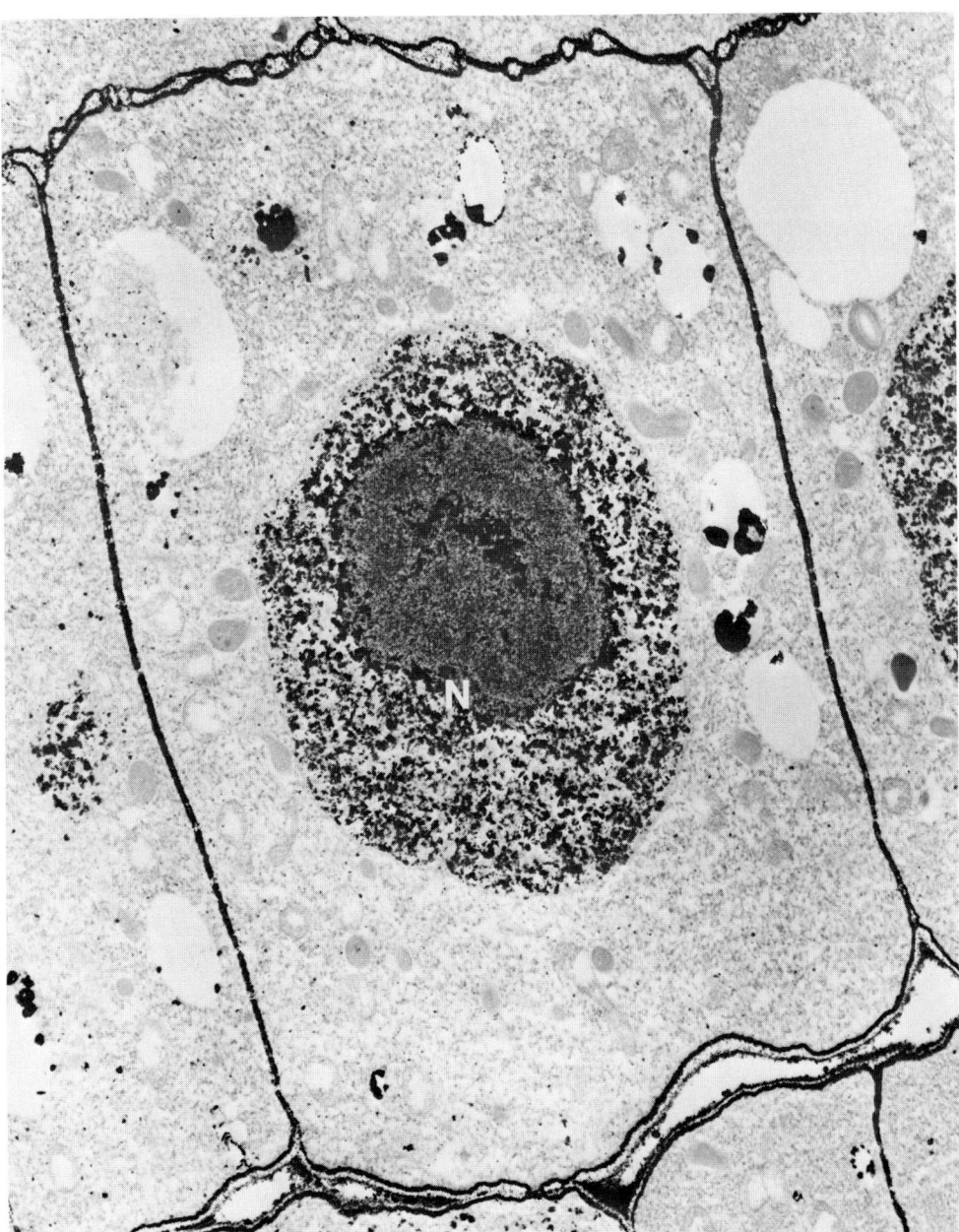

FIGURE 7. Cytochemical staining for acid phosphatase in a root cortical cell of *Zea mays*. Reaction product is located along the plasmalemma, in cell walls and vacuoles, and throughout the nucleus (N). (Magnification × 15,400.)

Controls should include (1) incubation of the tissue in a medium lacking substrate and (2) incubation in the presence of an inhibitor. Sodium fluoride (10 mM) has typically been used as a phosphatase inhibitor. However, Hall[21] has demonstrated that 10 mM sodium fluoride inhibits phosphatase activity by only 70%.

Another control involves treatments to deactivate the enzyme. Examples of such treatments include overnight incubation in 10% bufferred acrolein, immersing the tissue in boiling fixative for several minutes, and tryptic digestion.[8,22,23] Tryptic digestion is accomplished

Table 1
EFFECTS OF FIXATION ON THE ACTIVITY OF PHOSPHATASES IN PLANT TISSUES

Enzyme	Fixative	Concentration (%)	Time (hr)	Original activity after fixation (%)	Ref.
ATPase	Glutaraldehyde	1	2	45	13
	Glutaraldehyde	5	2	27	
	Formaldehyde	1	2	103	
	Formaldehyde	5	2	68	
Nucleoside diphosphatase	Glutaraldehyde	1	1.5	21	15
	Glutaraldehyde	4	1.5	19	
	Formaldehyde	1	1.5	92	
	Formaldehyde	4	1.5	88	
β-Glycerophosphatase	Glutaraldehyde	3	2	84	18
	Acrolein	10	2.5	24	

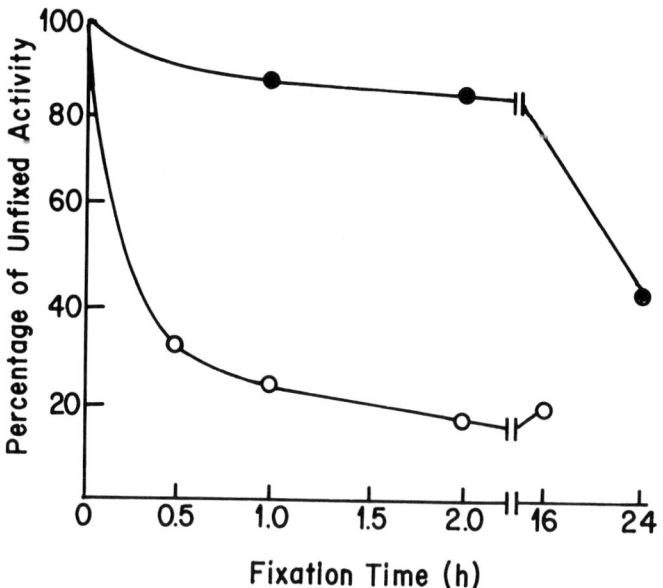

FIGURE 8. Effect of fixation on deactivation of phosphatases. Nucleoside diphosphatase from *Allium* roots (○) fixed in 4% glutaraldehyde (from Goff and Klohs[15]); β-glycerophosphatase from *Zea* roots (●) fixed in 3% glutaraldehyde. (From Sexton, R., Cronshaw, J., and Hall, J. L., *Protoplasma*, 73, 417, 1971. With permission.)

by incubating the tissue overnight in a mixture containing 1 mg trypsin in 100 mM Tris-HCl (pH 7.6) and 100 mM CaCl$_2$.[8]

Tissues should also be examined for lead phosphate absorption. This is accomplished by placing the tissues in 10 mℓ of 3.6 mM lead nitrate in 50 mM acetate buffer and, subsequently, adding 1 mℓ of 10 mM NaH$_2$PO$_4$. Tissues should then be washed and inspected for lead deposits.[8]

Table 2
EFFECT OF INCUBATION TIME ON DEPTH OF ACID PHOSPHATASE LOCALIZATION IN *ZEA MAYS* (FROM SEXTON AND HALL[8])

Incubation time (min)	Depth of Localization (μm)
4	60
120	90
240	100

Adapted from Sexton, R. and Hall, J. L., in *Electron Microscopy and Cytochemistry of Plant Cells*, Hall, J. L., Ed., Elsevier/North-Holland, Amsterdam, 1978, 63.

A TYPICAL PROTOCOL FOR PHOSPHATASE LOCALIZATION

The following protocol will serve as a good "starting point" for the cytochemical localization of phosphatase.[24] Details (e.g., composition of incubation medium) and modifications required for the localization of specific enzymes are presented in subsequent sections.

1. Fix tissues for 1 to 2 hr in a buffered aldehyde solution at temperatures not exceeding 4°C. Fixatives should be as pure as possible, since contaminants can inactivate phosphatases.[25] Do not use phosphate buffers.
2. Wash the tissue overnight at 4°C in buffer.
3. Section the tissue into slices less than 200 μm thick. Place tissue sections in a solution containing the same buffer and at the same pH as is to be used for the incubation. Tris-maleate buffer is preferred, since it stabilizes the lead nitrate and keeps it in solution.[7]
4. Place tissue sections in an incubation medium lacking substrate for 20 to 30 min at 4°C.
5. Place tissue sections in the complete incubation medium for times ranging from 0.5 to 2 hr. The method if mixing the incubation medium is important, since any precipitate of lead phosphate, once formed, is slow to dissolve. One method that will decrease chances of precipitate formation is to divide the buffer equally between two beakers. Add the substrate to one solution and add the lead nitrate to the other. Slowly mix the contents of the two beakers with rapid stirring.[24] Addition of 3% dextran to the incubation medium may also help prevent precipitation of lead.[26] Addition of polyvinylpyrrolidone (0 to 1 g/10 mℓ reaction mixture) may help provide an adequate colloidal osmotic pressure.[24]
6. Adequate incubation times can be estimated by placing the incubated tissue slices into 1% $(NH_4)_2S$ for 1 min and then observing them with a light microscope. If a black deposit is present, the incubation time is adequate.[8] Ericsson and Trump[27] recommend using $^2/_3$ of this light-microscope time for an ultrastructural localization.
7. Wash the tissue sections in a buffer solution for 20 to 60 min to remove excess incubation medium.
8. Dehydrate and embed the tissue using routine procedures. Section and observe the tissue unstained.

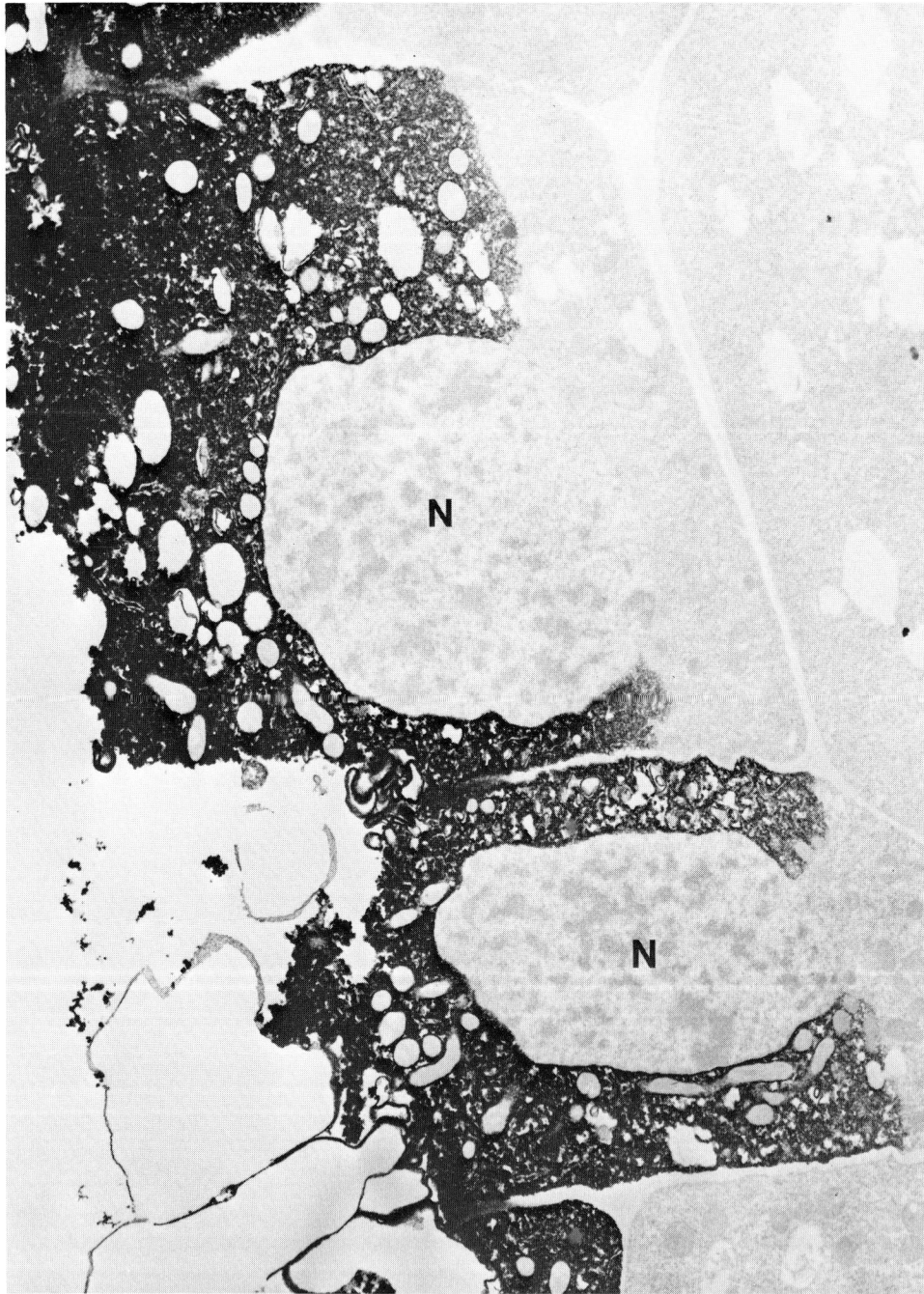

FIGURE 9. Poor penetration of substrate during the cytochemical localization of thiamine pytophosphatase (TPPase) in root cortical cells of *Zea mays*. Note the intensity of staining at the cut surface and the absence of staining in the innermost regions of the cells. N = nucleus. (Magnification × 13,300.)

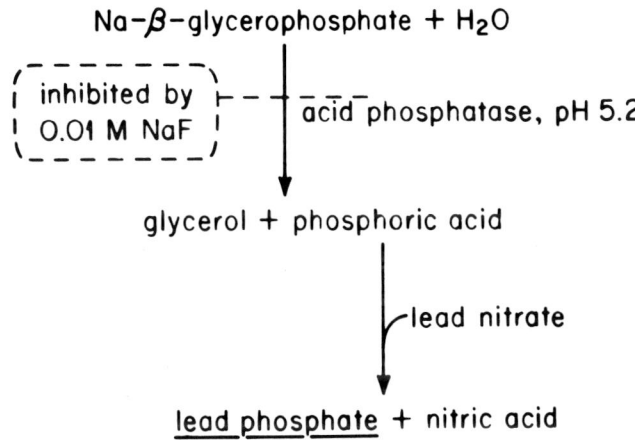

FIGURE 10. The lead-based technique for localizing acid phosphatase using sodium β-glycerophosphate as the substrate.

ACID PHOSPHATASES

Acid phosphatases refer to a group of enzymes having broad, but differing substrate specificities. Their pH optima range from 5.0 to 6.5. While some prefer the term acid phosphatase, others use more definitive names such as β-glycerophosphatase to describe the substrate used in their cytochemical localizations. However, this does not mean that β-glycerophosphatase attacks only β-glycerophosphate.

The cytochemical localization of acid phosphatase using sodium β-glycerophosphate as a substrate is presented in Figure 10. Other substrates utilized for cytochemically localizing acid phosphatase include *p*-nitrophenyl phosphate[28,29] and cytidine monophosphate.[30]

Protocol for Acid Phosphatase Localization[24]

1. Fix tissue in buffered glutaraldehyde for 2 hr at 4°C. Although inactivation of acid phosphatase is complete within 5 min,[9] longer fixation times are necessary to stabilize the tissue.
2. Wash tissue overnight at 4°C in a buffer other than phosphate. Tissues can usually be stored for several weeks at this stage without further loss of enzymatic activity.
3. Slice tissue into sections 50 μm thick and incubate the section for 1 hr in 0.05 M Tris-maleate buffer, pH 5.2. Poux[29,32] has recommended adding 10% sucrose in order to facilitate penetration and keep the medium isotonic with the cell.
4. Incubate the sections in a medium consisting of 40 mM Tris-maleate buffer (pH 5.2), 8 mM sodium β-glycerophosphate, and 2.4 mM lead nitrate for 10 to 60 min at room temperature. Some investigators[33] prefer to use this medium immediately after preparation, while others[17,27,32,34] store the medium (for up to 16 hr at 37°C) before use, apparently in response to the suggestion that this treatment reduces artifacts.[27]
5. Wash the tissue in the medium described in step 3 above. Poux[35-37] has suggested washing the tissue in citrate buffer (0.2 M, pH 4.8) to remove insoluble calcium salts (e.g., calcium phosphate), while Miller and Palade[38] recommend a 1-min rinse in buffer containing 4% formaldehyde.
6. Dehydrate and embed.

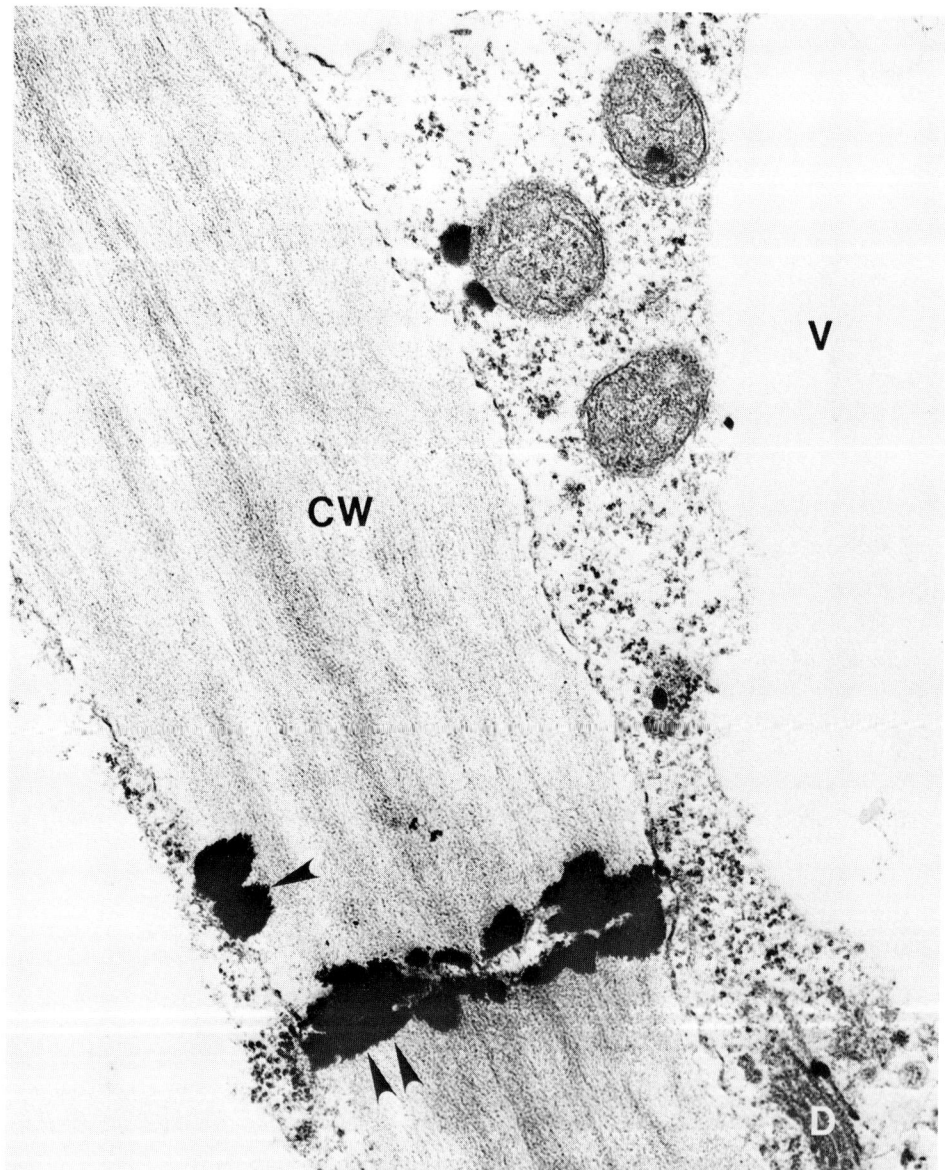

FIGURE 11. Distribution of acid phosphatase in an unwounded stem cortical cell of *Sedum telephoides*. Double arrows indicate reaction product at plasmalemma. CW = cell wall; D = dictyosome; V = vacuole. (Magnification × 39,100.) (From Moore, R. and Walker, D. B., *Protoplasma*, 109, 317, 1981. With permission.)

Localization and Function of Acid Phosphatases

Cytochemical localizations of acid phosphatase are presented in Figure 4, 7, 11, and 12. In cortical cells of *Zea* roots, acid phosphatase activity is cytochemically discernible in cells walls and, to a lesser degree, in vacuoles and (occasionally) nuclei (Figures 5 and 7). In internodal cells, enzymatic precipitate is associated with chloroplast membranes, dictyosomes, the tonoplast, mitochondrial membranes, and plasmodesmata (Figures 4 and 11).[39-42] Acid phosphatase has also been cytochemically localized in endoplasmic reticulum[43] and P-protein of sieve elements.[44]

The association of enzymatic activity with the plasmalemma and plasmodesmata (Figures 4 and 12) has prompted some investigators to suggest a role for acid phosphatase in inter-

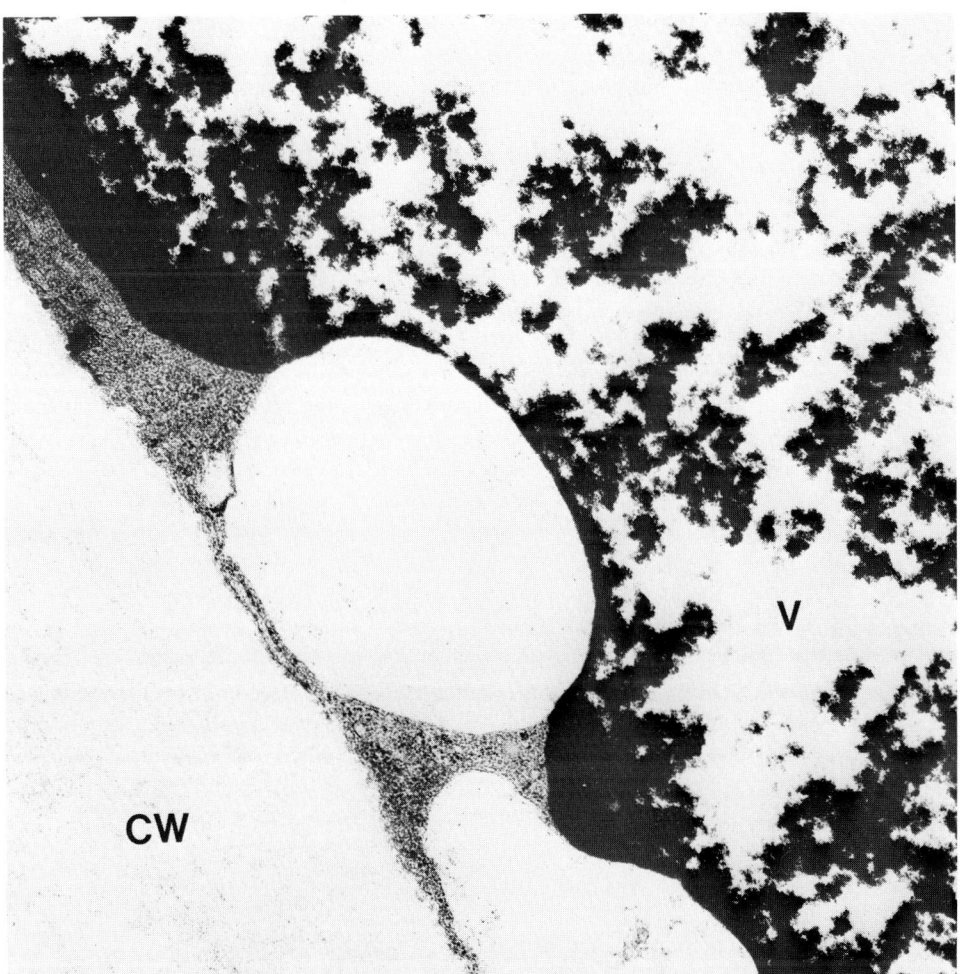

FIGURE 12. Distribution of acid phosphatase in a stem cortical cell of *Sedum telephoides* 7 days after wounding. Most enzymatic staining is associated with the tonoplast and central vacuole (V). CW = cell wall. (Magnification × 33,400.) (From Moore, R. and Walker, D. B., *Protoplasma*, 109, 317, 1981. With permission.)

cellular transport.[41,45,46] Acid phosphatase has also been suggested to be involved in the metabolism associated with seed germination,[47] autophagy during meiosis,[48] and cellular autolysis.[39,40,43]

Acid phosphatase is typically associated with cellular membranes and organelles and is absent from the cytosol in healthy cells.[39,42] Wounding results in a significant increase in the staining intensity associated with the tonoplast and vacuole (Figure 12).[39] Cellular necrosis correlates positively with the loss of this strict compartmentation of acid phosphatase activity in the cell (i.e., cytoplasmic release of the enzymes).[30] Gahan[49] believes that this release of acid phosphatase into the cytosol is the result, rather than the cause, of cellular necrosis. Interestingly, Brunk and Ericsson[50] have demonstrated that acid phosphatase can leak through "intact" membranes.

The cytochemical localization of acid phosphatase apparently varies depending on the substrate used for the localization. Oparka et al.[43] reported similar localization patterns when using either β-glycerophosphate or naphthol AS-BI phosphate as a substrate. However, a somewhat different localization was observed when *p*-nitrophenylphosphate was used as a substrate.

ADENOSINE TRIPHOSPHATASE (ATPase)

ATPases are nucleoside phosphatases that catalyze the dephosphorylation of ATP:

$$ATP \rightarrow ADP + Pi$$

ATPases describe a wide range of activities associated with different cellular locations and physiological functions, and include transport as well as coupling ATPases.[8] The range of phosphatase activities characteristic of many membranes and the nonquantitative nature of the procedure means that the precise specificity of cytochemical staining for ATPase is difficult to establish.[8]

Lead precipitation procedures for ATPase have been suggested to suffer from two problems: (1) lead inhibits ATPase and (2) lead catalyzes the nonenzymatic hydrolysis of ATP.[51-53] For example, Van Steveninck[54] suggested that membrane-associated deposits attributed to ATPase are nonspecific and contain no lead. However, Hall et al.[55] have shown that these membrane-localized deposits are of enzymatic origin and contain lead. Lead-based techniques have been defended by several other investigators[9,56,57] and have been applied to several plant systems.

Protocol for ATPase Localization

1. Fix tissue slices for a series of times ranging from 10 to 120 min in 1% glutaraldehyde in 0.05 M Tris-maleate (pH 7.2) at 4°C. ATPases are generally very sensitive to fixation and few of them will survive the procedures recommended for other phosphatases.
2. Wash the tissues overnight in cold buffer lacking fixative.
3. Incubate tissue slices 10 to 120 min in a reaction medium containing 4 mM Pb(NO$_3$)$_2$, 2.5 mM MgSO$_4$, 0.05 M Tris-maleate buffer (pH 7.2), and 3.5 mM ATP at room temperature
4. Rinse in buffer, dehydrate, and embed.

Additional Controls

Some ATPases are highly substrate specific. Thus, no staining should occur when ATP is replaced with other nucleotide phosphates such as GTP or ADP. Ouabain reportedly inhibits Na$^+$- and K$^+$-dependent ATPases.[58]

Localization and Function of ATPase

ATPases have been cytochemically localized in almost all cellular organelles,[34,47,59-61] an observation consistent with there being multiple forms of the enzyme.[62] Several investigators have noted a particularly intense staining of the plasmalemma,[44,63-68] prompting the suggestion that ATPase is involved with transport processes of the cell.[63,64,67] Typical cytochemical localizations of ATPase are shown in Figures 13 and 14.

OTHER NUCLEOSIDE PHOSPHATASES

The functions of nucleoside diphosphatases are largely unknown. However, they are important in electron microscopic cytochemistry, because they can be used to selectively stain various organelles.[24] Several nucleoside phosphatases have been localized in plant tissues. Aside from ATPase, the most extensively studied nucleoside phosphatase has been inosine diphosphatase (IDPase). IDPase is considered a marker for dictyosomes[8] and endoplasmic reticulum[9] in animal cells. Unlike ATPase, however, IDPase in fixed tissue is not inhibited by lead.[15]

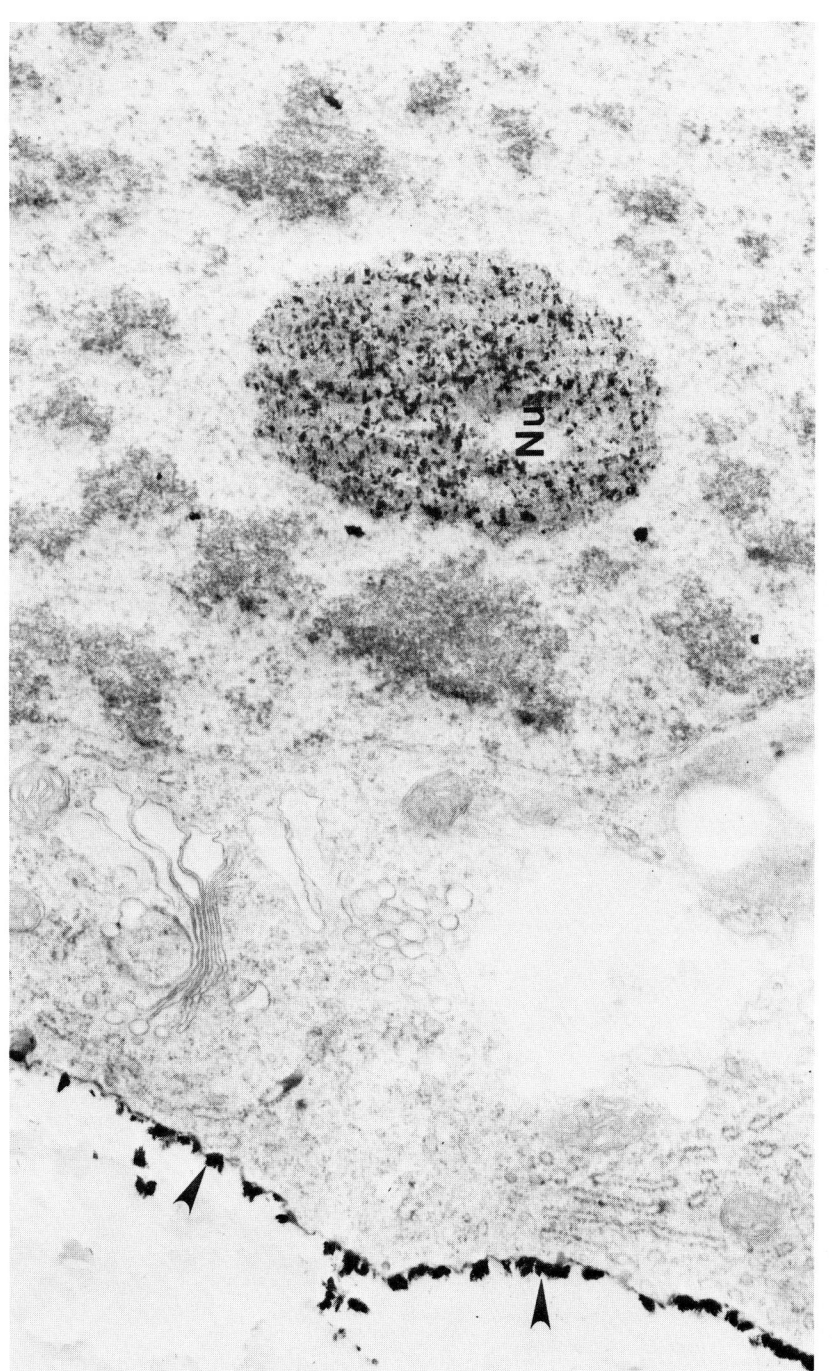

FIGURE 13. Cytochemical localization of ATPase in root cortical cells of *Zea mays*. Note enzymatic precipitate along plasmalemma (arrow) and in nucleolus (Nu). (Magnification × 27,700.)

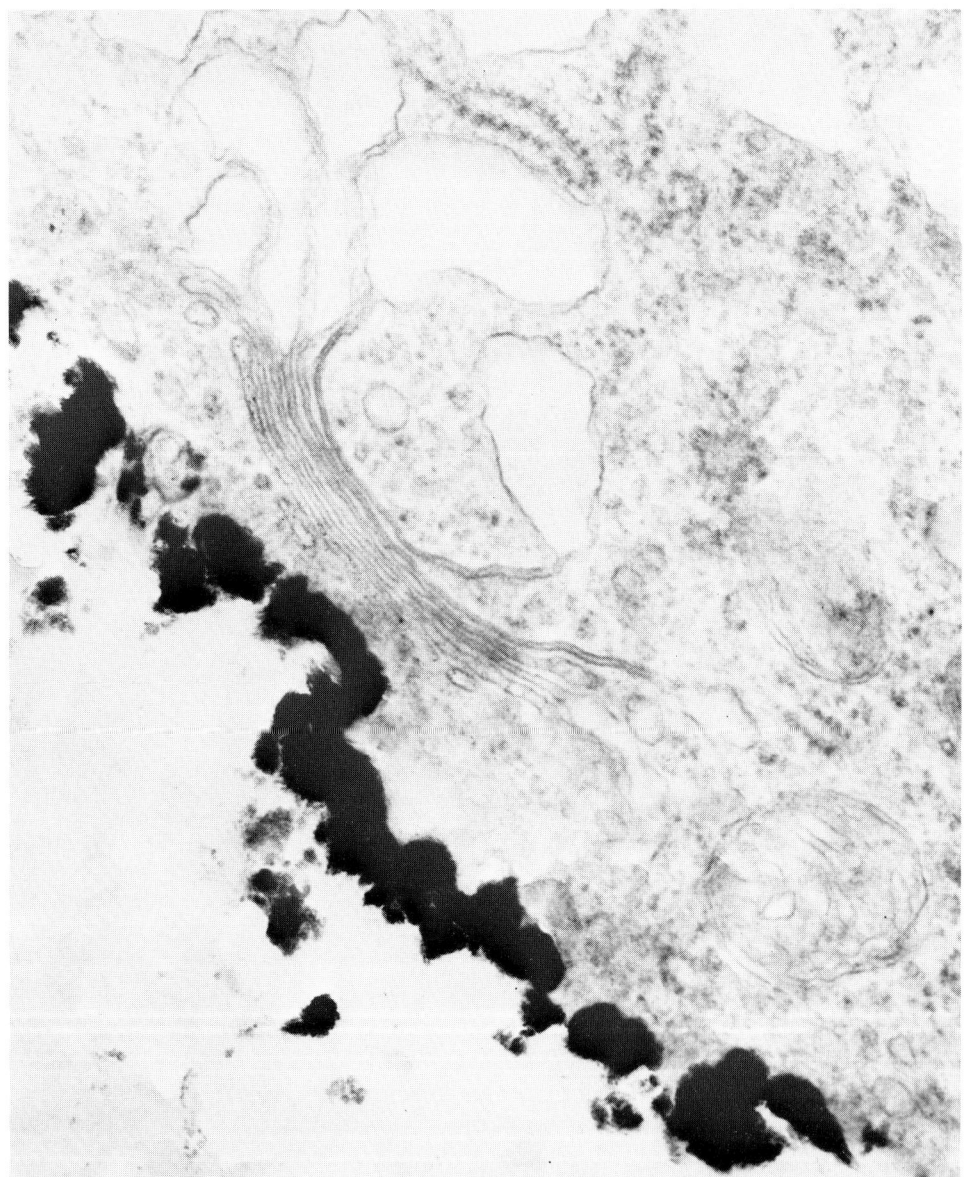

FIGURE 14. Cytochemical localization of ATPase along the plasmalemma of root cortical cells of *Zea mays*. (Magnification × 71,200.)

Protocol for IDPase Localization

1. Fix tissue slices for 10 to 60 min in buffered formaldehyde. IDPase is extremely sensitive to glutaraldehyde fixation.[15]
2. Wash in buffer.
3. Incubate the tissue slices in a medium containing 10 mM IDP, 5 mM MnCl$_2$, 2.4 mM Pb(NO$_3$)$_2$, and 50 mM Tris-maleate buffer (pH 7.35) for up to 2 hr at 37°C.
4. Rinse in buffer, dehydrate, and embed.

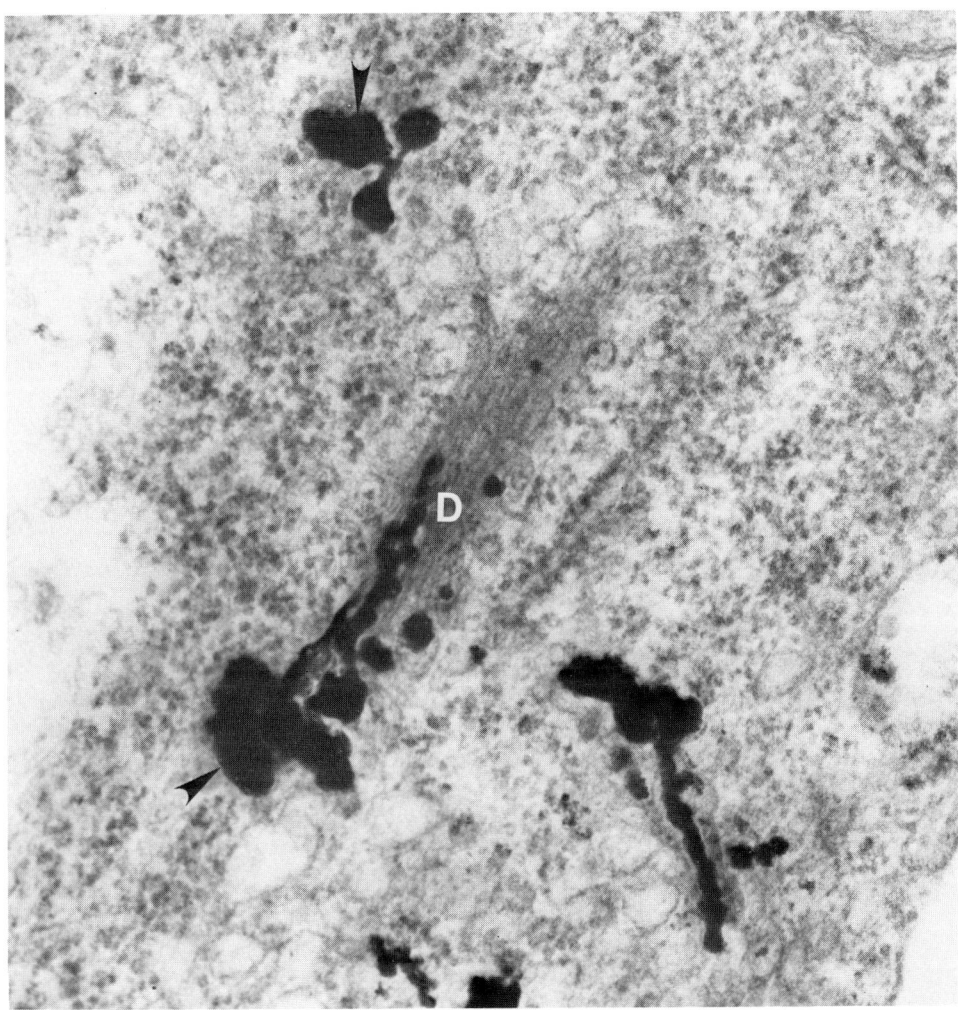

FIGURE 15. Cytochemical localization of inosine diphosphatase (IDPase) in a dictyosome (D) and its associated vesicles (arrows) in a peripheral cell of a root cap of *Zea mays*. (Magnification × 73,900.)

Localization of IDPase

Although typically reported to be specific for dictyosomes (and associated vesicles) (Figure 15) and endoplasmic reticulum (Figure 16), IDPase has also been localized cytochemically along the plasmalemma and vacuoles of plant cells.[17,37,69,70]

Protocol for Thiamine Pyrophosphatase (TPPase) Localization

1. Fix tissue for 2 to 4 hr at 4°C in glutaraldehyde.
2. Wash overnight in buffer.
3. Section tissue into slices 50 μm thick.
4. Incubate the sections in a medium containing 4 mM Pb(NO$_3$)$_2$, 5 mM MnCl$_2$, 80 mM Tris-maleate (pH 7.2), and 2.5 mM TPP (cocarboxylase) for 30 to 120 min at 37°C.
5. Rinse with buffer, dehydrate, and embed.

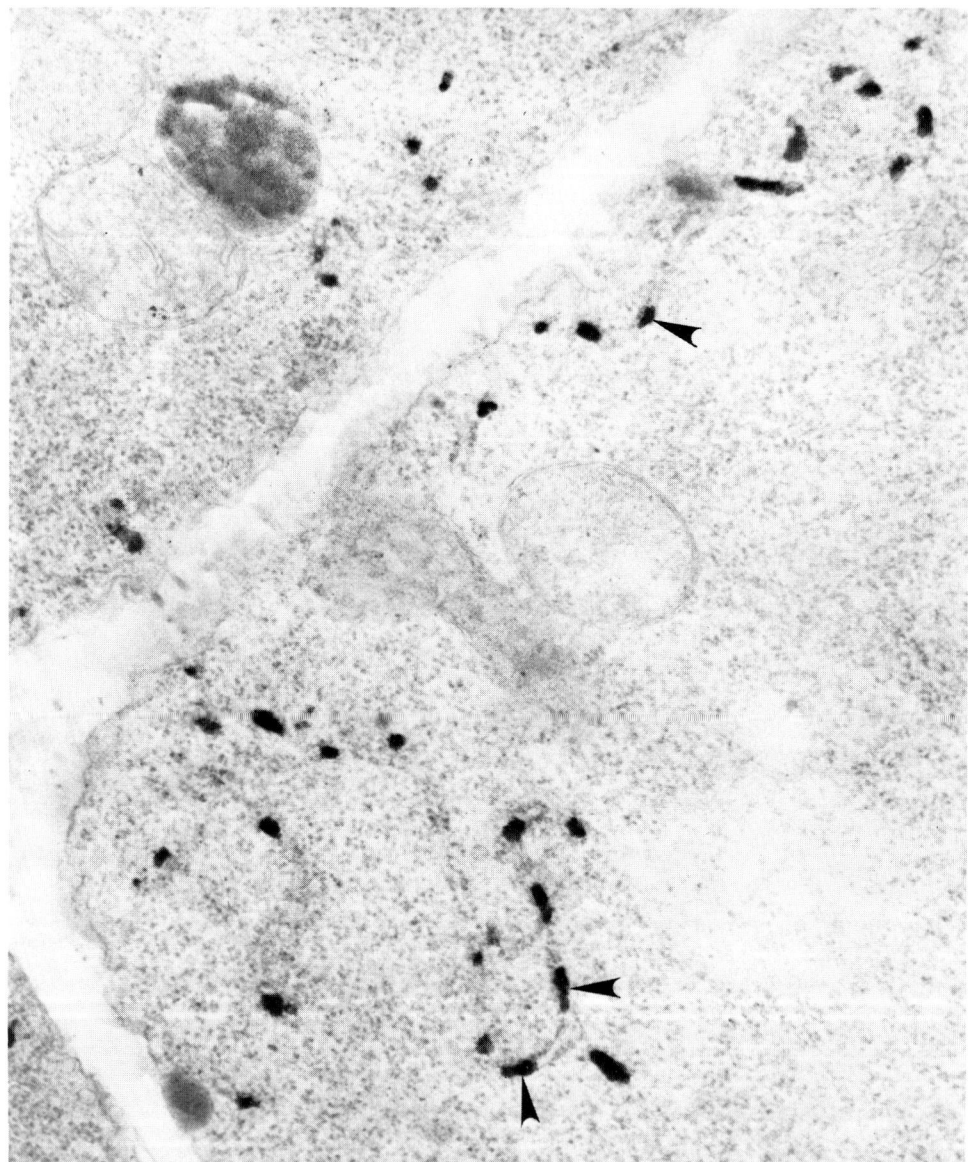

FIGURE 16. Cytochemical localization of IDPase in endoplasmic reticulum (arrows) of root cortical cells of *Zea mays*. (Magnification × 20,100.)

Additional Controls

Replace TPP with β-glycerophosphate (10 mM) or cytidylic acid (11.5 mg/10 mℓ reaction mix). Any staining with these substrates results from nonspecific phosphatases, but not TPPase.

Localization of TPPase

TPPase has been reported to be associated with dictyosomes,[17] the plasmalemma,[37] and vacuole.[37] Cytochemical staining of a cortical cell of *Zea mays* is shown in Figure 17. Note the enzymatic precipitate in the small vacuole and dictyosome (arrowhead).

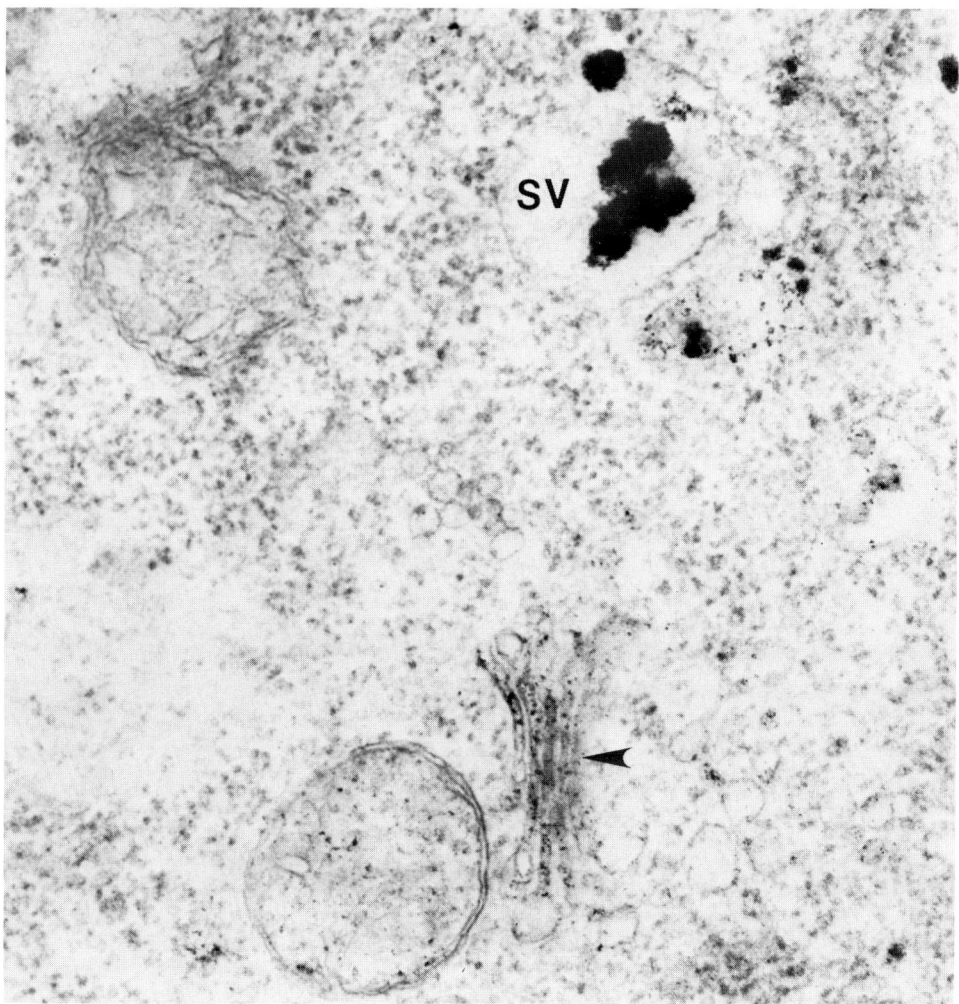

FIGURE 17. Cytochemical localization of thiamine pyrophosphatase (TPPase) in root cortical cells of *Zea mays*. Enzymatic precipitate is located in small vacuoles (SV) and dictyosomes (arrow). (Magnification × 59,000.)

ADENYLATE CYCLASE

Adenylate cyclase catalyzes the synthesis of 3′,5′-cyclic adenosine monophosphate (cAMP) from ATP:

$$ATP \rightarrow cAmp + PPi$$

The use of ATP as a substrate does not allow one to discriminate staining due to ATPase from that due to adenylate cyclase. However, 5′-adenylylimidodiphosphate (AMP-PNP) is a substrate specific for adenylate cyclase.[71,72]

$$AMP\text{-}PNP \rightarrow cAMP + \text{imidodiphosphate (PNP)}$$

AMP-PNP is apparently not hydrolyzed by membrane ATPases.[8]

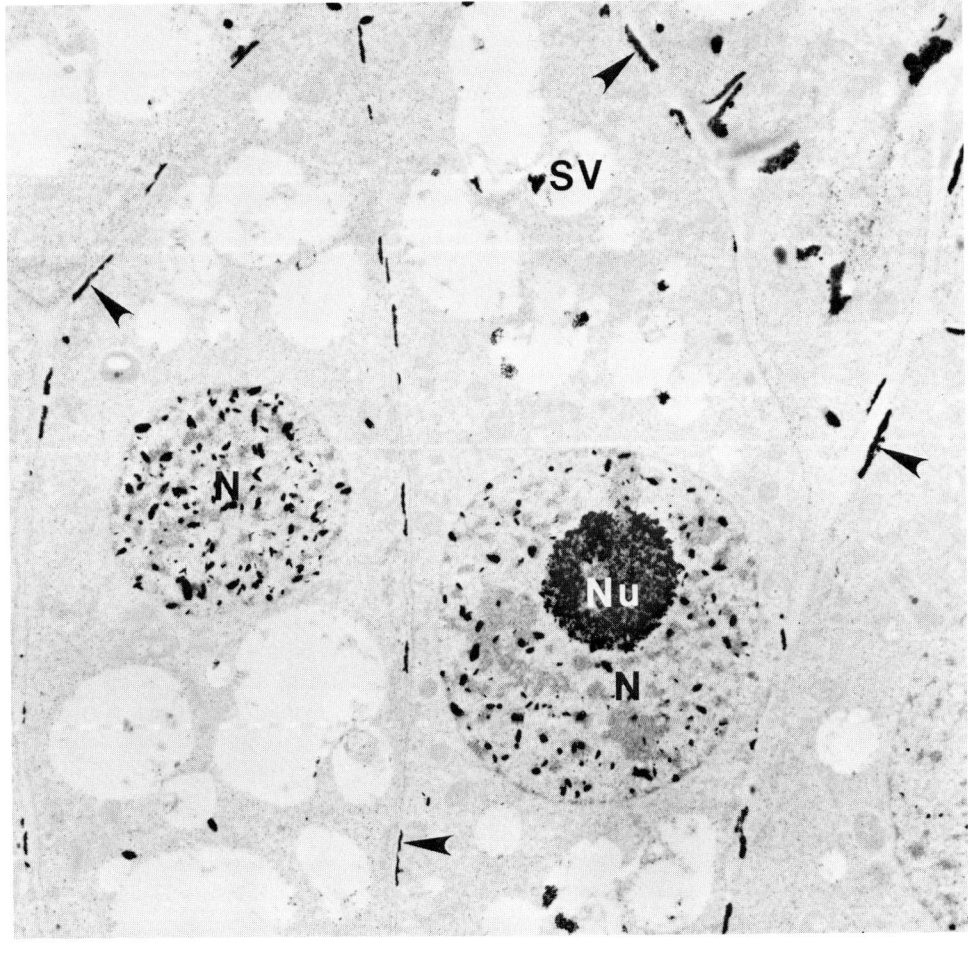

A

FIGURE 18. Cytochemical localization of adenylate cyclase in peripheral cells of root caps of *Zea mays*. Enzymatic precipitate was localized in the nucleus (N), nucleolus (Nu), small vacuoles (SV), and along the plasmalemma (arrows). (A) Magnification × 7890; (B) magnification × 16,400.

Protocol for Adenylate Cyclase Localization

1. Fix tissues 1 to 2 hr in buffered glutaraldehyde.
2. Wash overnight in buffer.
3. Incubate tissue slices for 30 to 60 min in a medium containing 0.5 mM AMP-PNP, 2 mM MgSO$_4$, 4 mM Pb(NO$_3$)$_2$, 8% dextran, and 80 mM Tris-maleate buffer (pH 7.4) at 30°C.
4. Rinse in buffer, dehydrate, and embed.

Localization and Function of Adenylate Cyclase

Adenylate cyclase is an enzyme responsible for the synthesis of cAMP, which is the active phosphorylating intermediate in many cellular control systems.[73] In cortical cells of *Z. mays*, adenylate cyclase is found primarily in the nucleolus, nucleus, vacuoles, and along the plasmalemma (Figures 18 and B). Adenylate cyclase has also been reported in the nuclear membrane and endoplasmic reticulum.[74]

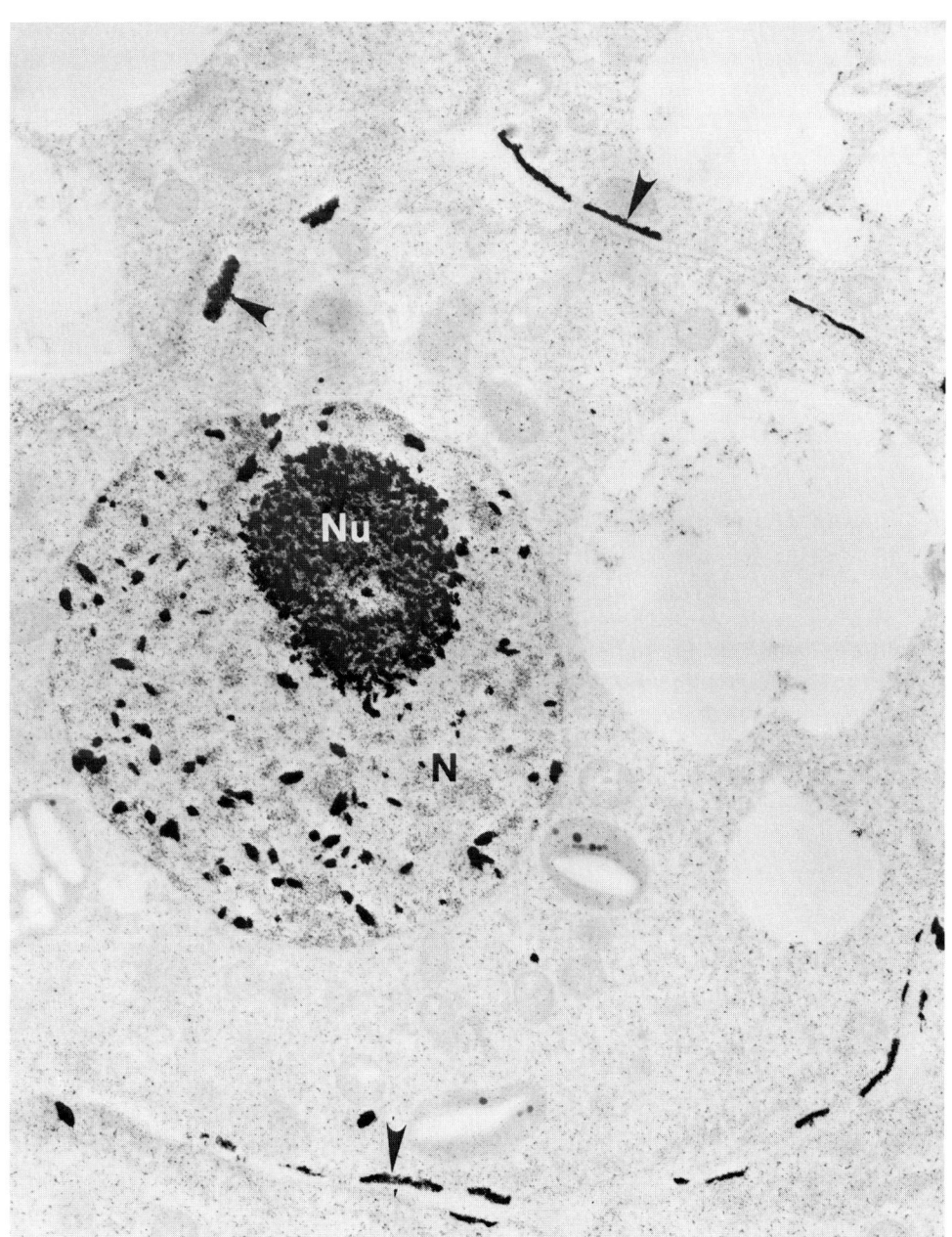

FIGURE 18B

GLUCOSE-6-PHOSPHATASE

Glucose-6-phosphatase is a magnesium-activated enzyme that catalyzes the dephosphorylation of glucose-6-phosphate:

$$\text{glucose-6-phosphate} \rightarrow \text{glucose} + \text{Pi}$$

Glucose-6-phosphatase can also hydrolyze several other hexose and pentose monosaccharides, such as fructose-6-phosphate and ribose-5-phosphate.[24] Glucose-6-phosphatase is very sensitive to aldehyde fixation.[75]

Protocol for Glucose-6-Phosphatase Localization[76]

1. Fix tissue 10 to 60 min in glutaraldehyde.
2. Wash the tissue overnight in buffer.
3. Incubate tissue slices in medium containing 60 mM Tris-maleate buffer (pH 6.5), 3 mM lead nitrate, and 4 mM glucose-6-phosphate for 30 to 60 min at 25°C.
4. Wash in buffer, dehydrate, and embed.

Additional Controls

1. Replace glucose-6-phosphate with glucose-1-phosphate, fructose-1,6-diphosphate, fructose-6-phosphate, ribose-5-phosphate, or β-glycerophosphate.[77]
2. Add Zn^{2+}, Cu^{2+}, Mg^{2+}, and Fe^{2+} ions to the incubation medium.[77,78]
3. Add D-glucose, ammonium molybdate, or the oral antidiabetic drug Orinase to the incubation medium.[79]

Localization and Function of Glucose-6-Phosphatase

Hall[80] has localized glucose-6-phosphatase along the plasmalemma and tonoplast of cells in roots of *Beta vulgaris*. The localization by McClelen and Moore[76] using root caps of *Z. mays* is similar to that reported by Hall,[80] but the localization changes during cellular differentiation. In columella (i.e., graviperceptive) cells located in the center of the cap, relatively small amounts of glucose-6-phosphatase staining are associated with the plasmalemma and cell wall (Figure 19). As columella cells differentiate into peripheral cells, enzymatic staining is associated with mucilage and, to a lesser extent, the cell wall (Figure 20). These observations prompted McClelen and Moore[76] to suggest that glucose-6-phosphatase is involved in the production and/or secretion of mucilage by root caps of *Z. mays*.

ALKALINE PHOSPHATASES

Alkaline phosphatases are nonspecific enzymes that can hydrolyze monoesters of phosphoric acid.[8] Their pH optima are between 7.6 and 9.9.[8] Alkaline phosphatases have been cytochemically localized using two different methods. In the "direct" method, the liberated phosphate is trapped by lead ions, usually in the presence of chelating agents to enhance lead solubility. In the two-step "indirect" method, the liberated phosphate is trapped by calcium, thus forming calcium phosphate. Calcium phosphate, which is insoluble at high pH, is then converted to a lead salt by washing the tissue in 2% lead nitrate.[24]

Alkaline phosphatases have not been localized cytochemically in plant tissues. Indeed, Sexton and Hall[8] have suggested that there is very little evidence for their presence in plants. Conversely, Yamaya and Matsumoto[81] have reported that alkaline phosphatase activity can be induced in cucumber roots by calcium starvation.

Protocol for Alkaline Phosphatase Localization

1. Fix the tissue in buffered glutaraldehyde. Alkaline phosphatase is very stable, and longer than normal fixation times should do no harm.[24]
2. Wash the tissue overnight in buffer.
3. Incubate tissue slices in a medium containing 3 mM cytidine monophosphate (CMP),

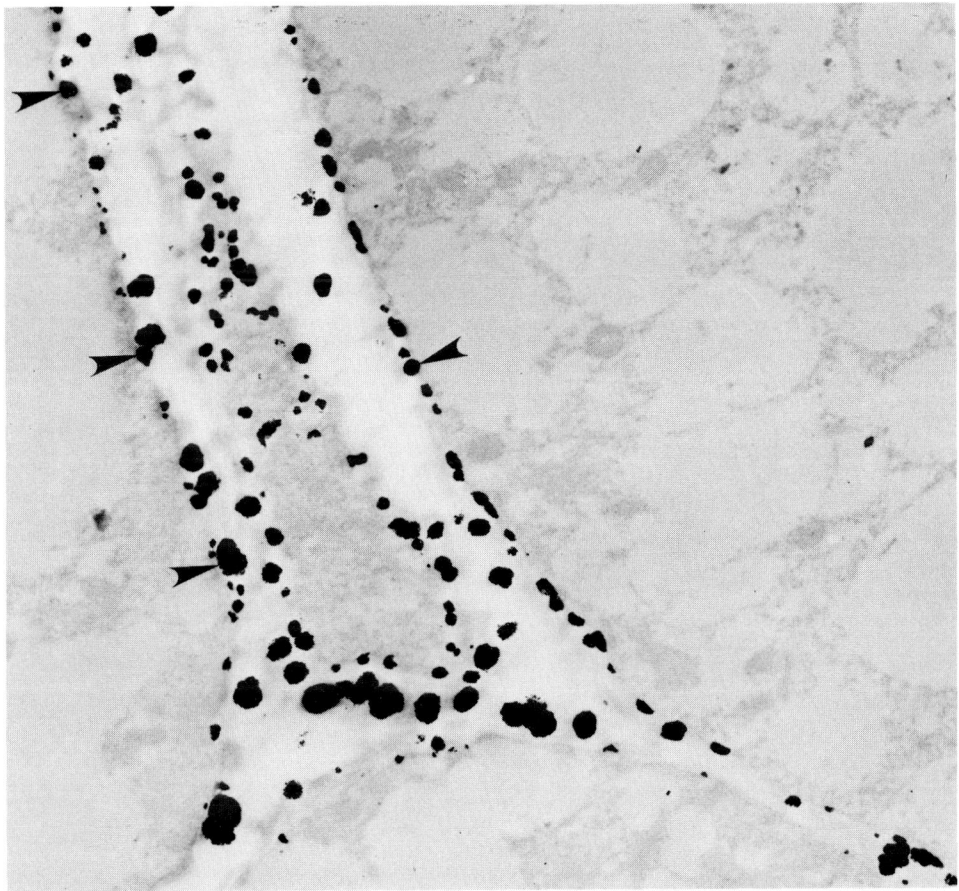

FIGURE 19. Cytochemical localization of glucose-6-phosphatase in columella cells of *Zea mays* roots. Arrows indicate reaction product along plasmalemma and cell wall. (Magnification × 20,100.)

    5 mM MgCl$_2$, 4 mM Pb(NO$_3$)$_2$, and 120 mM Tris-maleate, pH 8.5 at 4°C for 10 to 60 min.
4. Incubate other sections in the same medium having a pH of 7.2. Tissue must be processed at two pHs, since CMP is hydrolyzed by a variety of enzymes. Any staining revealed at pH 8.5, but absent at pH 7.2, is likely due to alkaline phosphatase.[24]
5. Rinse in buffer, dehydrate, and embed.

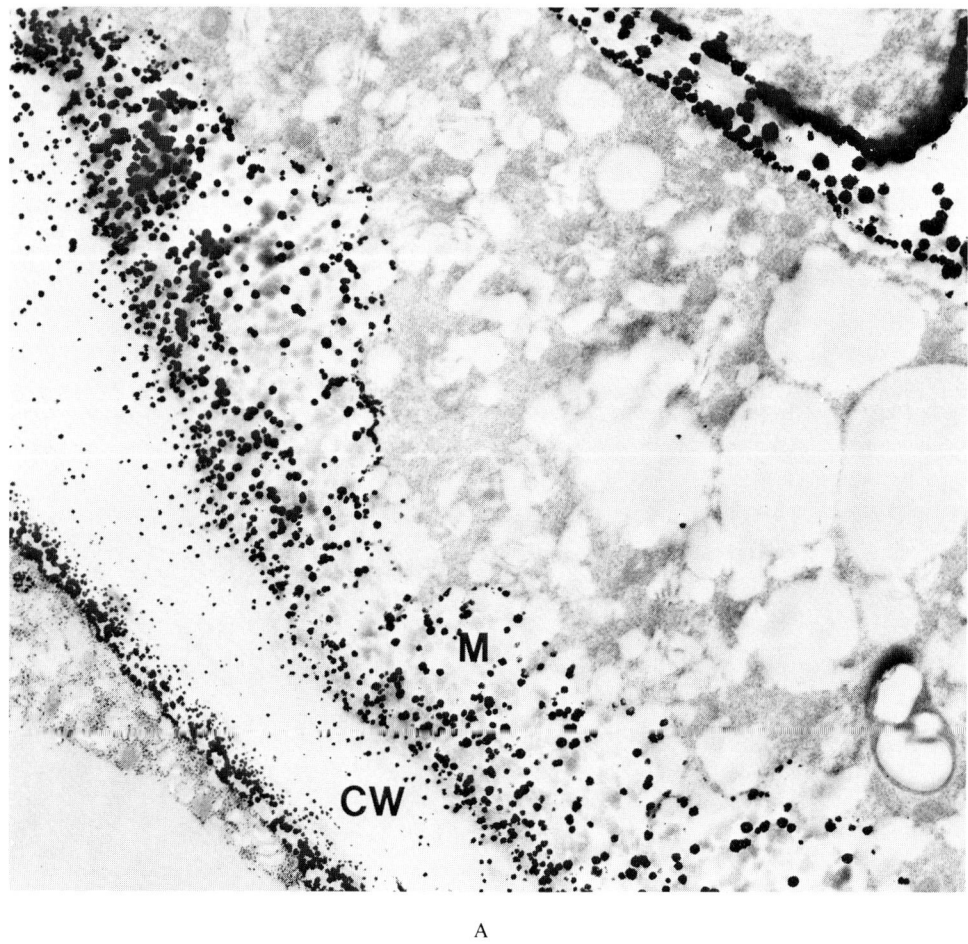

A

FIGURE 20. Cytochemical localization of glucose-6-phosphatase in mucilage (M) and cell walls (CW) in peripheral cells of root caps of *Zea mays*. (A) Magnification × 15,100; (B) magnification × 42,300.

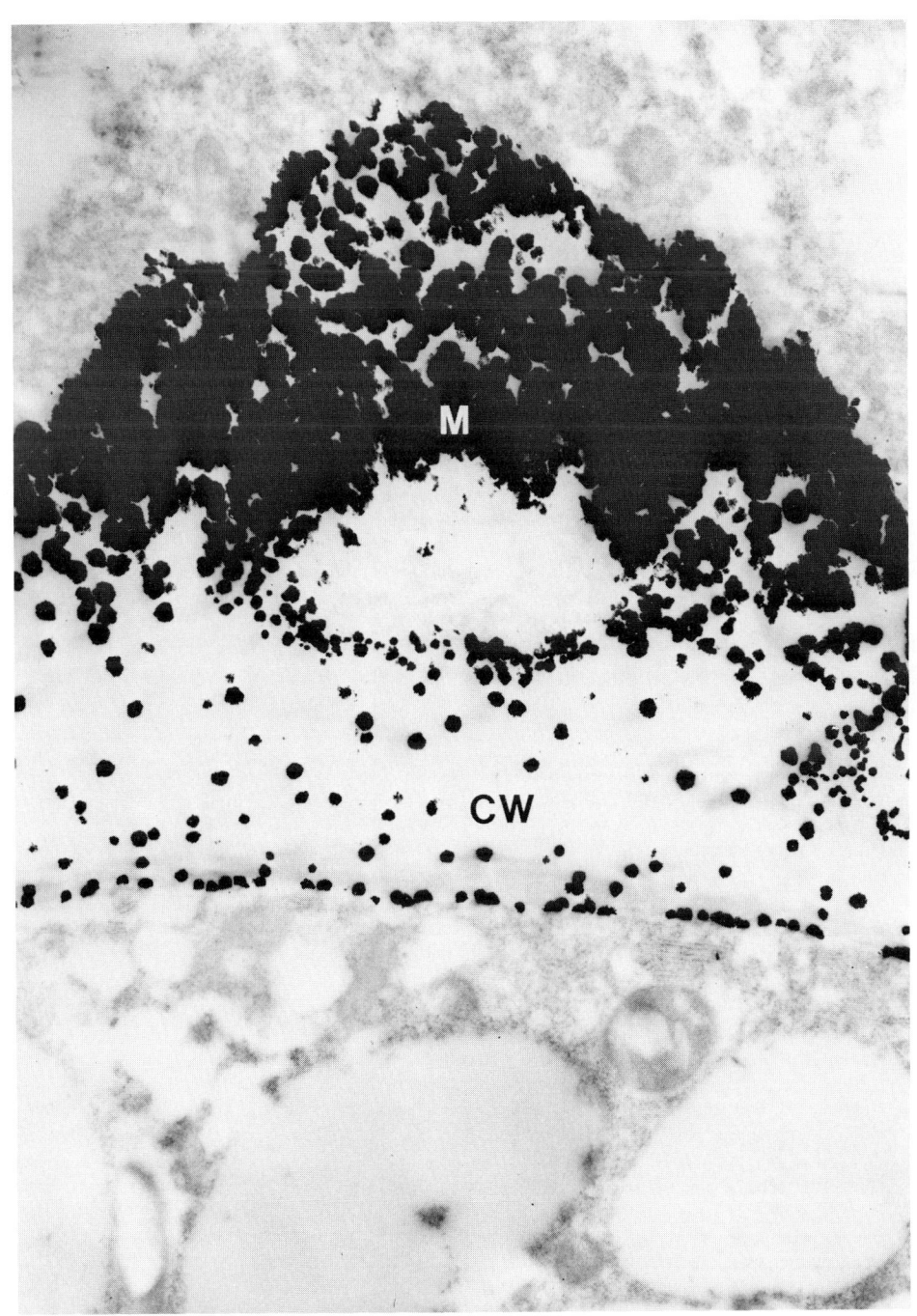

FIGURE 20B

REFERENCES

1. **Gomori, G.,** Microtechnical demonstration of phosphatase in tissue sections, *Proc. Soc. Exp. Biol. Med.,* 42, 23, 1939.
2. **Takamatsu, H.,** Histologische und biochemische Studien uber die Phosphatase, Histochemische Untersuchungsmethodik der Phosphatase und deren Vereilung in verschiedenen Organen und Geweben, *Acta Soc. Pathol. Jpn.,* 29, 429, 1939.
3. **Livingston, D. C., Coombes, M. M., Franks, L. M., Maggi, V., and Gahan, P. B.,** A lead phthalocyanin method for the demonstration of acid hydrolases in plant and animal tissues, *Histochemie,* 18, 48, 1969.
4. **Efron, Y.,** Specific inhibition of acid phosphatases in pollen of maize by the diazonium salt Fast Garnet GBC, *J. Histochem. Cytochem.,* 17, 734, 1969.
5. **Charvat, I. and Esau, K.,** An ultrastructural study of acid phosphatase localization in *Phaseolus vulgaris* xylem by the use of an azo-dye method, *J. Cell Sci.,* 19, 543, 1975.
6. **Bowen, I. D.,** A high resolution technique for the fine-structure localization of acid hydrolases, *J. Microsc. (Oxford),* 94, 25, 1971.
7. **Barka, T. and Anderson, P. J.,** Histochemical methods for acid phosphatase using hexazonium pararosanilin as coupler, *J. Histochem. Cytochem.,* 10, 741, 1962.
8. **Sexton, R. and Hall, J. L.,** Enzyme cytochemistry, in *Electron Microscopy and Cytochemistry of Plant Cells,* Hall, J. L., Ed., Elsevier/North-Holland, Amsterdam, 1978, 63.
9. **Essner, E.,** Phosphatases, in *Electron Microscopy of Enzymes,* Vol. 1, Hayat, M. A., Ed., Van Nostrand Reinhold, New York, 1973, 44.
10. **Cornelisse, C. J. and van Duijn, P.,** A new method for the investigation of the kinetics of the capture reaction in phosphatase cytochemistry. I. Theoretical aspects of the local formation of crystalline precipitates, *J. Histochem. Cytochem.,* 21, 607, 1973.
11. **Washitani, I. and Sato, S.,** On the reliability of the lead salt precipitation method of acid phosphatase localization in plant cells, *Protoplasma,* 89, 157, 1976.
12. **Hall, J. L. and Sexton, R.,** Cytochemical localization of peroxidase activity in root cells, *Planta,* 108, 103, 1972.
13. **Al-Azzawi, M. J. and Hall, J. L.,** Effects of aldehyde fixation on adenosine triphosphatases and peroxidase activities in maize root tips, *Ann. Bot.,* 41, 431, 1977.
14. **De Jong, D. W., Olson, A. C., and Jansen, E. F.,** Glutaraldehyde activation of nuclear acid phosphatase in cultured plant cells, *Science,* 155, 1672, 1967.
15. **Goff, C. W. and Klohs, W. D.,** Nucleoside diphosphatase in the onion root tip. I. Effects of fixation and lead on enzyme activity, *J. Histochem. Cytochem.,* 22, 945, 1974.
16. **Leskes, A., Siekevitz, P., and Palade, G. E.,** Differentiation of endoplasmic reticulum in hepatocytes. I. Glucose-6-phosphatase distribution *in situ, J. Cell Biol.,* 49, 264, 1971.
17. **Dauwalder, M., Whaley, W. G., and Kephart, J. E.,** Phosphatases and differentiation of the Golgi apparatus, *J. Cell Sci.,* 4, 455, 1969.
18. **Sexton, R., Cronshaw, J., and Hall, J. L.,** A study of the biochemistry and cytochemical localization of B-glycerophosphatase activity in root tips of maize and pea, *Protoplasma,* 73, 417, 1971.
19. **Coulomb, P. and Coulomb, C.,** Processes d'autophagie cellulaire, dan les cellules de meristemes radiculaires en etat d'anoxie, *C. R. Acad. Sci. (Paris),* 277, 1577, 1973.
20. **Coulomb, C. and Coulomb, P.,** Participation des structure golgiennes a la formation des vacuoles autolytiques at a leur approvisionnement enzymatique dans les cellules du meristeme radiculaire de la courge, *C. R. Acad. Sci. (Paris),* 277, 2685, 1973.
21. **Hall, J. L.,** Histochemical localization of B-glycerophorphatase acitivity in young root tips, *Ann. Bot.,* 33, 399, 1969.
22. **Nir, I. and Seligman, A. M.,** Ultrastructural localization of oxidase activities in corn root tip cells with two new osmiophilic reagents compared to diaminobenzidine, *J. Histochem. Cytochem.,* 19, 611, 1971.
23. **Odik, H.,** The reaction of mitochondria in the coleoptiles of rice (*Oryza sativa* L.) with diaminobenzidine, *J. Cell Sci.,* 17, 43, 1975.
24. **Lewis, P. R.,** Metal precipitation methods for hydrolytic enzymes, in *Staining Methods for Sectioned Material,* Lewis, P. R. and Knight, D. P., Eds., North-Holland, Amsterdam, 1978, 137.
25. **Anderson, P. J.,** Purification and quantitation of glutaraldehyde and its effect on several enzyme activities in skeletal muscle, *J. Histochem. Cytochem.,* 11, 652, 1967.
26. **Szmigielski, S.,** The use of dextran in phosphatase techniques employing lead salts, *J. Histochem. Cytochem.,* 19, 505, 1971.
27. **Ericsson, J. L. E. and Trump, B. F.,** Observations on the application to electron microscopy of the lead phosphate technique for the demonstration of acid phosphatase, *Histochemie,* 4, 470, 1965.
28. **Halperin, W.,** Ultrastructural localization of acid phosphatase in cultured cells of *Daucus carota, Planta,* 88, 91, 1969.

29. **Poux, N.**, Localisation d'activites enzymatiques dans le meristeme radicalaire de *Cucumis sativus* L. III. Activite phosphatasique acide, *J. Microsc. (Paris)*, 9, 407, 1970.
30. **Marty, F.**, Peroxisomes et compartiment lysosomal dans les cellules du meristeme radiculaire d'*Euphorbia characias* L.: une etude cytochimique, *C. R. Acad. Sci. (Paris)*, 273, 2504, 1971.
31. **Arborgh, B., Ericsson, J. L. E., and Helminen, H.**, Inhibition of renal acid phosphatase and aryl sulfatase activity by glutaraldehyde fixation, *J. Histochem. Cytochem.*, 19, 449, 1971.
32. **Poux, N.**, Localisation des activites phosphatasiques acides et peroxydasiques au niveau des ultrastructures vegetales, *J. Microsc. (Paris)*, 21, 265, 1974.
33. **Holt, S. J. and Hicks, R. M.**, The localization of acid phosphatase in rat liver cells as revealed by combined cytochemical staining and electron microscopy, *J. Biophys. Biochem. Cytol.*, 11, 47, 1961.
34. **Maier, K. and Maier, U.**, Localization of beta-glycerophosphatase and Mg^{++}-activated adenosine triphosphatase in moss haustorium, and the relation of these enzymes to the cell wall labyrinth, *Protoplasma*, 75, 91, 1972.
35. **Poux, N.**, Localisation des phosphatases et de la phosphatase acide dans les cellules der embryons de ble (*Triticum vulgare* Vill.) lors de la germination, *J. Microsc. (Paris)*, 2, 557, 1963.
36. **Poux, N.**, Localisation de l'activate phosphatasique acide et des phosphates dans les grains d'aleurone. I. Grains cristalloides, *J. Microsc. (Paris)*, 4, 771, 1965.
37. **Poux, N.**, Ultrastructural localization of aryl sulfatase activity in plant meristematic cells, *J. Histochem. Cytochem.*, 14, 932, 1967.
38. **Miller, F. and Palade, G. E.**, Lytic activities in renal protein absorption droplets: an electron microscopical cytochemical study, *J. Cell Biol.*, 23, 519, 1964.
39. **Moore, R. and Walker, D. B.**, Studies of vegetative compatibility-incompatibility in higher plants. III. The involvement of acid phosphatase in the lethal cellular senescence associated with an incompatible heterograft, *Protoplasma*, 109, 317, 1981.
40. **Matile, P.**, *The Lytic Compartment of Plant Cells*, Springer, Berlin, 1975, 183.
41. **Gahan, P. B. and McLean, J.**, Subcellular localization and possible functions of acid B-glycerophosphatases and naphthol esterases in plant cells, *Planta*, 89, 126, 1969.
42. **Moore, R.**, The cytochemical localization of acid phosphatase in plant cells, *Texas Soc. Electron Microsc. J.*, 13, 9, 1982.
43. **Oparka, K. J., Johnson, R. P. C., and Bowen, I. D.**, Sites of acid phosphatase in the differentiating root protophloem of *Nymphoides peltata* (S. G. Gmel.) O. Kuntze, *Plant Cell Environ.*, 4, 27, 1981.
44. **Catesson, A.**, Observations cytochimiques sur les tubes cribles de quelques angiospermes, *J. Microsc.*, 16, 95, 1973.
45. **Braun, H. J. and Sauter, H.**, Phosphatase lokalisation in phloembeckenzellen und siebrohren der Dioscoreaceae und ihre mogliche bedeutung fur den aktiven assimilat transport, *Planta*, 60, 543, 1964.
46. **Figier, J.**, Localisation infrastructural de la phosphomonoesterase acide dans la stipule de *Vicia faba* L. au niveau de nectaire, *Planta*, 83, 60, 1968.
47. **Akiyama, T. and Suzuki, H.**, Localization of acid phosphatases in aleurone layers of wheat seeds, *Z. Pflanzenphysiol.*, 101, 131, 1981.
48. **Schultz, P. and Jensen, W. A.**, Pre-fertilization ovule development in *Capsella*: ultrastructure and ultracytochemical localization of acid phosphatase in the meiocyte, *Protoplasma*, 107, 27, 1981.
49. **Gahan, P. B.**, A re-interpretation of the cytochemical evidence for acid phosphatase activity during cell death in xylem differentiation, *Ann. Bot.*, 42, 755, 1978.
50. **Brunk, U. T. and Ericsson, J. L. E.**, Cytochemical evidence for the leakage of acid phosphatase through ultrastructurally intact lysosomal membranes, *Histochem. J.*, 4, 479, 1972.
51. **Rosenthal, A. S., Moses, H. L., Beaver, D. L., and Schuffman, S. S.**, Lead ion and phosphatase histochemistry. I. Nonenzymatic hydrolysis of nucleoside phosphates by lead ion, *J. Histochem. Cytochem.*, 14, 698, 1966.
52. **Moses, H. L. and Rosenthal, A. S.**, Pitfalls in the use of lead ion for histochemical localization of nucleoside phosphatases, *J. Histochem. Cytochem.*, 16, 530, 1968.
53. **Tice, L. W.**, Lead-adenosine triphosphatase complexes in adenosine triphosphate histochemistry, *J. Histochem. Cytochem.*, 17, 85, 1969.
54. **Van Steveninck, R. F. M.**, The verification of cytochemical tests for ATPase activity in plant cells using X-ray microanalysis, *Protoplasma*, 99, 211, 1979.
55. **Hall, J. L., Browning, A. J., and Harvey, D. M. R.**, The validity of the lead precipitation technique for the localization of ATPase activity in plant cells, *Protoplasma*, 104, 193, 1980.
56. **Novikoff, A. B.**, Their phosphatase controversy: love's labours lost, *J. Histochem. Cytochem.*, 18, 916, 1970.
57. **Hall, J. L.**, Localisation of cell surface adenosine triphosphatase activity in maize roots, *Planta*, 85, 105, 1969.
58. **Borgers, M. and Thone, F.**, Further characterization of phosphatase activities using non-specific substrates, *Histochem. J.*, 8, 301, 1976.

59. **Hall, J. L.,** Cytochemical localization of ATPase activity in plant root cells, *J. Microsc. (Oxford),* 93, 219, 1971.
60. **Robards, A. W. and Kidwai, P.,** Cytochemical localization of phosphatase in differentiating secondary vascular cells, *Planta,* 87, 227, 1969.
61. **Price, G. D. and Whitecross, M. I.,** Cytochemical localization of ATPase activity on the plasmalemma of *Chara corallina, Protoplasma,* 116, 65, 1983.
62. **Edwards, M. L. and Hall, J. L.,** Intracellular localization of the multiple forms of ATPase activity in maize root tips, *Protoplasma,* 78, 321, 1973.
63. **Leigh, R. A., Williamson, F. A., and Wyn Jones, R. G.,** Presence of two different membrane-bound, KCl-stimulated adenosine triphosphatase activities in maize roots, *Plant Physiol.,* 55, 678, 1975.
64. **Leonard, R. T. and Van Der Woulde, W. J.,** Isolation of plasma membranes from corn roots by sucrose density gradient centrifugation: an anomalous effect of ficoll, *Plant Physiol.,* 57, 105, 1976.
65. **Cronshaw, J.,** ATPases in mature and differentiating phloem and xylem, *J. Histochem. Cytochem.,* 28, 375, 1980.
66. **Yapa, P. A. J. and Spanner, D. C.,** Localisation of adenosine triphosphatase activity in mature sieve elements of, *Tetragonia, Planta,* 117, 321, 1974.
67. **Bentwood, B. J. and Cronshaw, J.,** Cytochemical localization of adenosine triphosphatase in the phloem of *Pisum sativum* and its relation to the function of transfer cells, *Planta,* 140, 111, 1978.
68. **Browning, A. J., Hall, J. L., and Baker, D. A.,** Cytochemical localization of ATPase activity in phloem tissues of *Ricinus communis, Protoplasma,* 104, 55, 1980.
69. **Goff, C. W.,** Localization of nucleoside diphosphatase in the onion root tip, *Protoplasma,* 78, 397, 1973.
70. **Zaar, K. and Schnepf, E.,** Membranflub and nucleosiddiphosphatase-reaktion in Wurzelhaaren von *Lepidium sativum, Planta,* 88, 224, 1969.
71. **Rodbell, M., Birnbaumer, L., Pohl, S. L., and Kranz, H. M. J.,** The glucagon-sensitive adenyl cyclase system in plasma membranes of rat liver, *J. Biol. Chem.,* 246, 1877, 1971.
72. **Howell, S. L. and Whitfield, M.,** Cytochemical localization of adenyl cyclase activity in rat islets of langerhans, *J. Histochem. Cytochem.,* 20, 873, 1972.
73. **Bitensky, M. W. and Gorman, R. E.,** Cellular responses to cyclic AMP, *Prog. Biophys. Mol. Biol.,* 26, 409, 1973.
74. **Al-Azzawi, M. J. and Hall, J. L.,** Cytochemical localization of adenyl cyclase activity in maize roots, *Plant Sci. Lett.,* 6, 285, 1976.
75. **Shnitka, T. K. and Seligman, A. M.,** Ultrastructural localization of enzymes, *Ann. Rev. Biochem.,* 40, 375, 1971.
76. **McClelen, C. E. and Moore, R.,** The cytochemical localization of glucose-6-phosphatase in plant cells, *Tex. Soc. Electron Microsc. J.,* 15, 11, 1984.
77. **Tice, L. W. and Barrnett, R. J.,** The fine structural localization of glucose-6-phosphatase in rat liver, *J. Histochem. Cytochem.,* 10, 754, 1962.
78. **Chiquoine, A. D.,** The distribution of glucose-6-phosphatase in the liver and kidney of the mouse, *J. Histochem. Cytochem.,* 1, 429, 1953.
79. **Mahler, H. R. and Cordes, E. H.,** *Biological Chemistry,* Harper & Row, New York, 1971, 497.
80. **Hall, J. L.,** Fine structural and cytochemical changes occurring in beet discs in response to washing, *New Phytol.,* 79, 559, 1977.
81. **Yamaya, T. and Matsumoto, H.,** Properties of alkaline phosphatase in cucumber roots induced by calcium starvation, *Plant Cell Physiol.,* 22, 1355, 1981.

DEHYDROGENASES

M. Ekés

GENERAL CONSIDERATIONS

Dehydrogenases belong to the first class of the enzymes catalyzing all sorts of oxido-reduction reactions in the cells (oxido-reductases). These enzymes carry out oxidation by transferring hydrogen from organic substrates (donors) to natural hydrogen (electron) acceptors, but not to molecular oxygen.[1] They can be divided into several groups.[2-4]

Pyridine (nicotinamide) nucleotide (NAD [P])-linked dehydrogenases (e.g., malate dehydrogenase [MDH]) and glyceraldehyde 3-phosphate dehydrogenase (GAPDH) have low redox potentials (about 320 mV) and transfer two electrons and proton(s) to the reversibly dissociable coenzymes NAD(P). A few hundred dehydrogenases belong to this group.

The enzymes of the second group will oxidize the reduced coenzymes NAD(P)H. These NAD(P)H-oxidizing enzymes have covalently linked flavin prosthetic groups and are of intermediate redox potentials (about −50 mV). They are often named alternatively as NAD(P)H-dehydrogenases, NAD(P)H-tetrazolium or ferricyanide (oxido) reductases, diaphorases, etc., since many pathways for the oxidation of reduced coenzymes have been described and their exact identification is as yet uncertain. The majority of these enzymes are extramitochondrial; they are present in the cell both in soluble and membrane-bound state.

The third group is constituted by flavoprotein dehydrogenases, e.g., succinate dehydrogenase (SDH) (redox potential + 30 mV), which transfers reducing equivalents from substrates, not to NAD(P), but directly to another acceptor present in the electron transport chain of mitochondria or in another compartment.

The principle of the cytochemical localization of dehydrogenases is based on the ability of certain chemical compounds, e.g., tellurite,[5] tetrazolium salts, or ferricyanide, to act as artificial (one- or two-) electron acceptors which, when reduced, produce, either directly or indirectly, a colored insoluble precipitate.

The site of action of an artificial electron acceptor is determined by many factors, mainly by its redox potential, but properties such as lipid solubility or the ability to act as a one- or two-electron acceptor might have decisive influence on the actual site of action in given circumstances.[6] It is generally accepted that only flavoprotein oxidoreductases, e.g., SDH, monoamine oxidase (MAO), and the large amount of the various NAD(P)H-oxidizing enzymes containing FMN or FAD in their prosthetic groups, are able to react directly with the usual artificial electron acceptors.[7] Therefore, one has to bear in mind that in case of examination of other dehydrogenases the reduction of the electron acceptor takes place through the activity of endogenous (flavoprotein) diaphorase(s). The final precipitate, although primarily depending on the activity of the enzyme under study, will localize the sites of activity of the diaphorase(s).[4,8-11]

From among the many hundreds of dehydrogenases, only a minority has been fully characterized biochemically and even less of them has been localized to a particular intracellular site. The situation is even more so with plant dehydrogenases, the main difficulties being caused by the peculiarities of the plant cells (weak permeability, presence of cell wall, and, in most cases, large vacuoles) and by the usually extreme sensitivity of these enzymes to fixatives. As a consequence of methodical difficulties, ultracytochemistry of plant dehydrogenases is a largely neglected sphere of investigation, and very few publications have appeared so far.[12-20]

In this review the emphasis will predominantly be laid on the rationale of the effects and by-effects of the electron acceptors (tetrazoles, ferricyanide) and of the components of the ferricyanide-containing incubation medium, with some suggestions for their reasonable utilization in plant dehydrogenase ultracytochemistry. Certain limitations in the interpretability of some of the controls involved in the experiments for ultrastructural demonstration of malate- and succinate-driven ferricyanide oxidoreductase activities[12-15] will similarly be stressed. Some remarks will also be made on certain controversial points in connection with the advantages and disadvantages of tissue pretreatment (prefixation, freezing, etc.) prior to proper incubation vs. incubation of native samples. In addition, besides surveying some of the media and methods used most frequently in light microscopic dehydrogenase cytochemistry, full descriptions are given of the very few methods for ultrastructural examination of altogether only a couple of dehydrogenases that have been tried so far. For more details on principles and practice of dehydrogenase cytochemistry, the reader is advised to consult several basic works on this topic that is covered as a whole in the fundamental book of Wohlrab et al.[11] but much useful information on the subject can also be found in the basic works of Pearse[21] and Lojda et al.[4] on general histochemistry. Several aspects of the basic principles and a wide range of experimental procedures, used with animal materials for dehydrogenase ultracytochemistry, are given in the book of Geyer[22] and in the review of Hanker,[3] but for methods in conjunction with more control experiments the reader is referred rather to the methodical repertory of Totovič.[2] Unfortunately, almost no summarizing work on plant dehydrogenase cytochemistry exists. Jensen's classical *Botanical Histochemistry*[8] similarly, to the more recent *Botanical Microtechnique and Cytochemistry* by Berlyn and Miksche,[23] devoted merely a small part of a chapter to light microscopic demonstration of dehydrogenase (by the use of tetrazoles). In the recent excellent review of Sexton and Hall[74] on the present state and methods of plant ultrastructural cytochemistry, for dehydrogenases, inevitably a rather short section only could have been allocated. Notwithstanding the tempting plenty of methods of dehydrogenase ultracytochemistry, one should not forget that in the overwhelming majority of cases they were elaborated for animal tissues. Uncritical and direct application of these methods to plant materials might easily lead to erroneous conclusions.

THE ELECTRON ACCEPTORS

Tetrazolium Salts

Hundreds of tetrazolium salts have been synthesized since 1894, yet only a few of them could have been used with success in dehydrogenase cytochemistry.[3,11,22] Although the ease with which a given tetrazole can be reduced largely determines whether it will be suitable for cytochemical purposes, its practical applicability is strongly influenced by other factors as well. The majority of the most commonly used tetrazolium salts is sufficiently soluble in water to be present in optimal concentration of 0.25 to 1 mg/mℓ of the incubation medium. Difficulties might arise, however, with the use of some ditetrazoles with large particle sizes (e.g., Nitro-blue tetrazolium chloride [NBT] or Tetra-nitro BT [TNBT]), because of their very slow diffusion through semipermeable membranes. Lipid-soluble tetrazoles (e.g., dimethylthyazolyl tetrazolium bromide [MTT], a monotetrazole) penetrate easier through lipoprotein membranes, but their accumulation at these sites or in lipid droplets might lead to aspecific formazan production. Artifactual localization of the final reaction product might also result from aspecific adsorption to proteins and lipoproteins (substantivity). Another important factor is the concentration-dependent enzyme-inhibitory effect of various tetrazoles.

Requirements for the formazan reduction product of the tetrazolium salts are similarly rather strict. First of all, they should have arisen as a direct or indirect result of the activity of the enzyme under study. The product should be intensely colored (for light microscopy) or electron-dense (for electron microscopy), preferably noncrystalline, with small particle sizes permitting finer localization. It should not migrate from the site of precipitation during incubation and processing for light or electron microscopic examination, enduring inorganic and organic solvents, embedding procedures, electron beam, etc.

Product substantivity to proteins is considered to be advantageous for prevention of product diffusion, unless this substantivity appears solely for certain cell components. Diformazans (e.g., that of NBT), owing to their long and flat molecules, can attach to protein molecules by hydrogen linkages if no interfering substituent groups are present. Monoformazans (e.g., of MTT) do not possess substantivity for proteins. Besides their substantivity properties, weaker solubility in lipids and organic solvents renders diformazans more suitable for cytochemical purposes. Of the two tetrazoles, NBT and TNBT,[25] employed presently in plant histochemistry, NBT is used predominantly, in spite of the obvious advantages of TNBT, evident from the comparison of their properties:[2-4]

1. Although the water solubility of NBT is better than that of TNBT, the penetration of the very large molecules of NBT into cells might be rather aggravated. Higher lipid solubility of NBT might give rise to artifactual precipitation.
2. TNBT accepts electrons more easily.
3. TNBT possesses better substantivity properties.
4. Adsorption of the diformazan of TNBT to lipids in the cell is considerably less than that of NBT diformazan, decreasing, hence, the probability of aspecific precipitation. Diformazan from NBT has a tendency for crystallization, is liable to coalesce into droplets in some preparations, and might form relatively large granules on lipid-water interfaces.
5. The granule size of TNBT diformazan precipitate is remarkably smaller (diameter of *circa* 50 Å as opposed to 300 to 900 Å of NBT diformazan). Nevertheless, even this size appears too large for high-resolution examinations.
6. TNBT diformazan is relatively more stable under the electron beam of the electron microscope.
7. TNBT diformazan has a somewhat less solubility in reagents used for electron microscopic dehydration and embedding.

Owing to its rather disadvantageous properties for electron microscopy, NBT has been employed, almost without exception,[20,26] only for light microscopic examination of dehydrogenases. Potentialities of newer tetrazoles[2,3,11,22] are not yet exploited in plant cytochemistry.

The reduction of the tetrazolium salts in mitochondria, according to the conception of Seidler,[27] happens uniformly at the flavoprotein-ubiquinone level and not at different regions of the respiratory chain ("reduction sites"), as was supposed earlier.[28] Reduction proceeds stepwise through flavin enzyme radicals and tetrazole radicals. In the presence of oxygen tetrazoles are reduced, as a matter of fact, nonenzymatically, by the superoxide radical arising from the reaction of the (enzymatically) reduced flavin enzyme (and/or ubiquinone or phenazine methosulfate) with molecular oxygen. Under aerobic conditions the relationship of the redox systems tetrazolium salt (tetrazolium radical and oxygen) superoxide radical will practically be decisive, whether the tetrazolium salt will or will not be able to accept electrons.

The validity of the concept of the superoxide anion radical (and superoxide dismutase) playing a central role in the aerobic tetrazolium reduction is usually not denied. Nevertheless, as was shown by Ponti et al.,[29] NBT was reduced anaerobically at about the same rate as aerobically. These observations warn against premature conclusions when interpreting the mechanism of NBT reduction patterns in tissue sections or cell suspensions possessing heterogenous oxygen tension. Data of the histochemical observations using tetrazoles, as a whole, are in good agreement with those obtained in biochemical experiments. Sites of increased metabolic activity, as judged from intense formazan production due to dehydrogenase activity, could be detected, e.g., in phloem companion and parenchyma cells,[30,31] in Strasburger cells,[32] and in transfer cells.[33,34] Marked changes in the enzyme activity distribution among the various tissues of differentiating roots were indicative of intense metabolic processes.[35] The considerable heterogeneity in the enzyme distribution of the mitochondria, observed by Avers and Tkal,[26] was later visualized ultracytochemically as well.[12,15] Observations using tetrazoles revealed, however, at the same time the serious limitations of the histochemical approach,[35] e.g., the real danger of artifactual localization of NAD(P)-linked dehydrogenases, connected partly with the presence of a firmly bound and a soluble diaphorase system in the cells.[36,37]

Ferricyanide

The principle of the use of ferricyanide for ultrastructural enzyme cytochemistry came from the "direct coloring" method of Karnovsky and Roots[38] devised for the visualization of acetyl choline esterase activity in 1964. In this method ferricyanide is reduced to ferrocyanide by the thiol groups liberated from the artificial substrate acetyl thiocholine as a result of enzymic hydrolysis. The primary reaction product, ferrocyanide, which is soluble, is captured by copper ions present in the incubation medium to yield an insoluble amorphous electron-opaque deposit (final reaction product) at the site of enzyme activity. The water- and lipid-insoluble precipitate, forming very fine granules (40 to 80 vs. 30° Å of NBT) of sufficient electron density, eliminated the main drawbacks of the tetrazolium salts and offered a very potent tool for ultrastructural enzyme cytochemistry.

The original principle of capturing enzymatically produced thiol groups was used for the demonstration of nonspecific esterase and acid phosphatase,[38] for localization of malate synthase in the glyoxysomes of cucumber and sunflower cotyledons,[40,41] and was modified for the localization of acyltransferases.[42,43] The potentialities of the ferricyanide were tried, also, for the ultracytochemical localization of PS II in chloroplasts.[44]

Most progress with the use of ferricyanide as electron acceptor has been made in the ultracytochemistry of dehydrogenases.[3,11,22] The breakthrough came with the successful application of the method of Karnovsky and Roots[38] for the ultracytochemical demonstration of SDH and other dehydrogenases by Ogawa et al.[45,46] and Kerpel-Fronius and Hajós.[47] The widespread use of ferricyanide has undoubtedly been promoted by the fact that this compound, owing to its high positive redox potential (+430 mV), has since 1938 been employed as an artificial one-electron acceptor in many biochemical assays acting as terminal electron acceptor.[6,11] According to biochemical[48,49] and ultracytochemical[46,47] observations, ferricyanide at higher concentrations (1 to 1.5 mM) directly accepts electrons on the flavoprotein level (and at ubiquinone[50] in mitochondria). The concentration of ferricyanide has, therefore, influence on the position of its reduction in the respiratory chain,[48,51] and using higher concentrations seems preferable for the direct localization of mitochondrial as well as the many extramitochondrial flavoprotein dehydrogenases demonstrated biochemically in a wide variety of sources. Plasmalemma-bound ferricyanide reductases (or NADH oxidases) from animal,[52] plant,[20,53,54] and fungal[55] materials may function as a generator of proton gradient across the plasma membrane and may play a role in transport processes. The role of the NADH cytochrome c reductase associated with a membrane fraction from protein bodies[56] or of the various NAD(P)H dehydrogenases from seeds[57] is yet unknown. These and other NAD(P)H oxidizing enzymes may function among others as an initial step in a redox chain either alone or as a part of a diaphorase, e.g., complexed with nitrate reductase. The widespread distribution of NADH-ferricyanide reductases among endomembranes (endoplasmic reticulum, dictyosomes, vesicles) led to the conception that NADH oxidoreductases are intrinsic membrane proteins "that may be conserved and/or modified during transfer from endoplasmic reticulum to the plasma membrane" as was stated by Morré et al.[58] Apart from several attempts[12-15,20] the exact methodical conditions for ultracytochemical demonstration and identification of NAD(P)H oxidoreductases in plant tissues have not yet been elaborated.

Besides the vast advantages of the ferricyanide as electron acceptor, its use in cytochemical assays might, nevertheless, involve some problems as well.[6,59] One is the possibility of nonenzymic formation of electron-dense deposits with electrons, accepted by ferricyanide nonenzymatically from certain redox dyes,[6] or thiol and other groups that become accessible during unfolding of proteins. Denaturation, rise in pH, temperature, and the presence of copper ions accelerate the rate of these reactions. By oxidizing functional thiol groups certain enzymes might be inhibited. Ferricyanide can directly oxidize, also, NADH, but the rate is slow. In other instances unexpected effects might ensue from the fact that ferricyanide is a large negatively charged anion: repelled by some negatively charged groups of protein it might be unable to gain access to enzyme and no electron transfer would occur between them. On the other hand, interaction with the substrate-binding cationic groups of an enzyme (e.g., SDH) might result in inhibition to some extent.

Ferricyanide (0.5 mM) irreversibly inhibits electron transport between water and PS II in pea chloroplasts,[60] and several enzymes of glycolysis and other enzymes might also be adversely affected by a millimolar concentration of ferricyanide.[59]

In cytochemical practice, however, major difficulties are connected with the nonpenetrating character of ferricyanide.[11-15,19,49,58,61,62] It is not able to penetrate through the intact inner membrane of mitochondria (a feature advantageously employed, however, for detecting biochemically the sidedness of membrane-bound dehydrogenases)[49] and cannot act as a Hill reagent with chloroplasts possessing intact inner envelope membrane. As a consequence of the presence of permeability barriers in the plasmalemma (and in the cell interior), ferricyanide is very poorly diffusible into tissues. In spite of the damage caused by tissue preparation, by some ingredient(s) of the medium and by the incubation process itself, reacted cells can usually be found only in a narrow surface zone of the samples.[12-15,47]

An interesting consequence of the nonpermeant character of ferricyanide, giving food for meditation, is that while ferricyanide cannot permeate the intact dissociation product HCN rapidly goes through it. When 0.3 mM ferricyanide was added to a growing *Chlorella* culture, as was reported by Pistorius,[63] the small amount of HCN formed was quite sufficient for complete elimination of nitrate uptake and inactivation of cellular nitrate reductase.

COUPLING (PRECIPITATING) AGENT, CHELATOR

Cupric (Cu^{2+}) ions are used for precipitation of the soluble primary reaction product in the described method. In order to prevent them from the formation of insoluble copper hydroxide and precipitation with unreacted ferricyanide, cupric ions must be kept chelated in the incubating solution.[47,51] The choice of this chelator is of crucial importance. For a precise localization of the enzyme activity the rapid saturation of the incubating medium by copper ferrocyanide at the site of its formation is essential. This process is determined by the velocities of both the reduction of ferricyanide and the dissociation of Cu^{2+} chelate complex. "Precipitation of the (primary) reaction product cannot occur before the saturation of the environment (at least that closest to the level of reaction) with ferrocyanide up to a concentration exceeding the dissolubility threshold of copper ferrocyanide."[51]

As opposed to tartrate,[47] citrate[38,45,46] as a complexing agent has unfavorable properties: the slow release of copper from the copper citrate complex supposedly due to the kinetic stability of the latter does not permit a sufficiently rapid formation of copper ferrocyanide, causing, thereby, a product diffusion (at pH 7.0). X-ray microanalysis of the final precipitate led Weavers[64] to have the opinion that it does not arise from a simultaneous coupling reaction (Hatchett's brown),[3,6,46,47] but ferrocyanide is initially deposited at the sites of reductase activity, the copper being, subsequently, attracted to form copper ferrocyanide complexes. Precipitation of ferrocyanide can be achieved with the use of other coupling agents as well: uranyl acetate and manganous chloride[42,43] eliminate the necessity of a chelator, since they do not form any precipitate with the compounds of the incubation medium. It cannot be excluded that the elimination of the chelator from the medium, besides resulting in less complex kinetics of the precipitate formation, would offer more possibility to cope with some problems of plant dehydrogenase ultracytochemistry.

THE INCUBATION MEDIUM

Precipitating Agent (Cu^{2+}), Chelator (Tartrate) and the Electron Acceptor (Ferricyanide) in the Incubation Medium

As a general rule, it is noteworthy that in order to obtain a clear medium using higher concentrations of $CuSO_4$ (21 mM), the molar concentration ratio of tartrate to $CuSO_4$ must be higher (>8.1) than in cases when the concentration of $CuSO_4$ is lower, 14 mM (>5.7) or 7 mM (1.4). The tendency to obtain a clear medium is enhanced when ferricyanide is present in higher concentrations (1.5 or 1 vs. 0.7 mM), that is favorable for *in situ* localization of flavoprotein enzymes, even when the concentration of the chelator is low.[51] The ferricyanide-containing medium should be a balanced system; it allows, however, some variation in the amount and proportions of its components. Precipitation in the medium of Kerpel-Fronius and Hajós[47] for SDH activity did not take place unless the ratio of tartrate (300 mM) to $CuSO_4$ (21 mM) was lowered from 14.3:1 below 10:1; 10 mM copper instead of the usual 21 mM appeared to be sufficient to deposit all amount of the ferrocyanide produced. Diffusion phenomena arose only when the molarity of the tartrate was doubled. Ferricyanide in lower concentration (0.05 vs. 1.5 mM) yielded a less intense reaction, while elevated concentrations of it neither produced more deposit nor caused diffusion.[47] For plant materials the use of 150 mM tartrate to 15 mM cupric ions and 1.5 mM ferricyanide seemed to be advantageous in the experiments carried out by Ekés,[14,15] since, as a rule, improved tissue penetration and better ultrastructural preservation could be achieved.

Substrates

Organic acids, substrates of many dehydrogenases, are added into the medium preferably in Na-salt form. Acids (e.g., malic acid) should previously be neutralized by NaOH and buffer with the use of a pH meter, since acidification of the medium beyond certain limits may adversely affect the enzyme activity and/or upset the cytochemical reaction. Moreover, strong pH shifts might cause precipitation in the medium rendering it unsuitable for the use. In order to be able to demonstrate the substrate availability, substrate should not become a limiting factor in the cytochemical assay. The usual concentration of the substrates rarely exceeds 140 mM. Elevated amounts of substrates might, however, result in too intensive accumulation of the reaction product rendering the site of its origin indiscernible.[12] Restrained penetration of certain substrates might be promoted by preincubation of the samples for a short time with the substrate in a phosphate-buffered sucrose solution, incubation under mild vacuum and/or at lower pH, or by mild disruption of the samples with sonication.[17] There are substrates capable occasionally of contributing to the chelating of the precipitation agent and enhancing, thereby, the stability of the medium;[51] that cannot always be considered as advantageous. NADH or NADPH is added usually in the amount of 0.5 to 1 mg/mℓ of the medium; NAD or NADP 0.2 to 1 mg/mℓ of the medium.

Employing these soluble coenzymes is unfortunately not exempt from some problems. Impurities in the NADH preparations can sometimes lead to precipitation in the medium. However, even with dissolved NADH lack of positive reaction might occur unexpectedly in some instances. Reduced coenzymes are unstable in weak acidic solutions, and although the process is rather slow, they might undergo structural modification and/or destruction even at pH 7.0 catalyzed by some components of buffer solutions.[66] NAD and NADP, on the other hand, are relatively unstable at pH 7.4 and have a tendency to condense with certain substances.[4] Absence of reaction may be caused, also, by diminished penetration through permeability barriers and the possibility of detrimental alterations in coenzyme structure by nucleotidases inside the cell cannot sometimes be ruled out. Moreover, conversion of the added coenzyme into the other in the tissues might also render the evaluation of the observations more difficult, e.g., added NADP or NADPH can be converted by endogenous phosphatase into NAD and NADH, respectively. In addition, large quantities of the coenzymes are inevitably removed from their places and washed out during cytochemical procedures. In order to counterbalance the many factors reducing the amount of available coenzymes under cytochemical conditions, it is advisable to maintain their high level by the addition of sufficient NAD(P)H (for NAD[P]H oxidoreductase reactions) or sufficient NAD(P) for conversion to NAD(P)H (for NAD[P]-linked dehydrogenase reactions).[21] The danger involved in interpreting the localization of NAD(P)-linked dehydrogenases should be kept in mind. Freely moving NAD(P) present in the cells and added with the incubation medium might equally be reduced by these diffusible enzymes both within and outside the tissue sample. Reduced coenzymes are reoxidized by tissue-bound and soluble NAD(P)H oxidoreductases (diaphorases), and for this reason the distribution of the reaction product will reflect the site of these NAD(P)H oxidoreductase(s) rather than that of the NAD(P)-linked dehydrogenases.

Finally, taking into consideration the interconnections of the metabolic pathways and the large amount of endogenous substrates in plant cells, one cannot always decide with full certainty whether the observed precipitate resulted from the direct or indirect action of the added substrate.[27]

pH of the Medium, the Buffer, and the Osmolarity

It has to be emphasized that since the pH optima of the precipitation of the functioning of the enzyme in the complex milieu of the cell and in isolated state might be very different, the chosen final pH of the medium must be based on a compromise. Moreover, this value should not exceed pH 8 for avoiding dissolution of copper ferrocyanide.[47,51] Phosphate buffers (0.05 to 0.15 M, pH 6.9 to 7.6) proved to be most suitable for dehydrogenase histochemistry. In the elaborated routine procedure for demonstration of SDH-, MDH-, and NADH-ferricyanide oxidoreductase activities in plant materials, a medium with pH 7.0 was employed with the use of 0.1 M K-Na-phosphate buffer, pH 7.6,[14,15] although media of pH values 6.7 to 7.2 caused no appreciable differences in precipitate distribution. Lower pH should be preferred with regard to easier penetration of the substrates into the tissues.[65] Structural damages caused partially by the unbalanced differences in the osmotic pressures of the unfixed tissues and of the incubation medium could have only insignificantly been reduced by addition of 1% (w/v) sucrose to the medium used by Valanne.[16] Addition of sucrose (200, 100, and 50 mM) into the medium for the routine procedure, though it enhanced its stability,[51] had no appreciable advantages and was not subsequently practiced.

Duration and Temperature of the Incubation

Periods of incubation at room temperature (20 to 40 min) were routinely used in the experiments of Ekés.[12-15] In materials incubated for as short a period as 5 min the reacted zone rarely spread beyond the mechanically damaged outer cell layer. Although more or less specific deposition did occur in these cells, the probability of aspecific precipitation and/or a lack of reaction due to enzyme inhibition to leakage of coenzymes and of soluble enzymes is increased. Meanwhile, more slowly reacting sites may remain unnoticed. Prolonged incubations, on the other hand, lead to excessive accumulation of product, blurring the initial sites of precipitation and, due to it, no differences between the strongly and weakly reacted sites can be detected.[3,67] In addition, the danger of artifactual precipitation and of structural deterioration grows with prolongation of the incubation time. Incubation at elevated temperature (37°C) as compared to that at room temperature yielded higher amount of precipitate without changing the subcellular distribution, but often resulted in more damaged ultrastructure. Neither higher temperature nor prolonged incubation improved, however, the rather limited permeability of unfixed plant tissues.

Phenazine Methosulfate (PMS)

As was noted before, under usual histochemical conditions (below pH 8) NAD(P)H cannot donate electrons directly to artificial acceptors. As a consequence of a possible weak activity and/or small amount of endogenous acceptor-reductases in the immediate vicinity of NAD(P)-linked dehydrogenase activity, not all of the produced diffusible NAD(P)H can be captured and reduced further at the sites of its origin. This might easily lead to artifactual precipitation and to loss of reducing equivalents. The use of certain soluble, low molecular weight redox intermediators such as phenazine methosulfate (PMS),[3,4] 1-methoxy PMS,[68] meldola blue (MB),[69-71] or methylene blue[72] has frequently increased the amount of demonstrable NAD(P)-linked dehydrogenases. The effect of PMS is usually explained by its ability to catalyze the direct transfer of electrons from NAD(P)H to tetrazoles, by-passing the tetrazolium reductases. According to Raap et al.,[7] however, PMS in small enzymatic sites (such as those in cells) functions rather as stimulator of these NAD(P)H oxidizing flavoproteins and as an efficient acceptor of reducing equivalents. Reduced $PMSH_2$, having low water solubility and good substantivity, does not diffuse away and tetrazolium reduction will occur at the site of PMS reduction. Notwithstanding, some loss of reducing equivalents, mainly when demonstrating NAD(P)-linked dehydrogenases with lower concentrations of tetrazoles, cannot be excluded.[69] The stimulatory effect of PMS is not always wanted and using PMS in lower concentrations is recommended.[11]

Other disadvantages of this flavin-like redox dye are its possible inhibitory effect on the histochemical demonstration of certain dehydrogenases, its reducibility by endogenous substrates, and that $PMSH_2$ is autooxidizable. Moreover, by-passing tetrazolium salts, it may transfer electrons to cytochrome oxidase that would diminish the amount of formazans. Meldola blue (MB), when tried in a system for quantitative SDH histochemistry,[71] resulted in SDH activity only 23% of that found with PMS applied, but in other experiments[69] MB was as efficient as PMS for the activity of this enzyme. However, a decisive advantage of MB is its much lower sensitivity to light. The effectiveness of a given redox intermediator is largely dependent on the peculiarities of the tissues under study[72] and is influenced by many other factors as well. When applied for light microscopic dehydrogenase histochemistry of plant materials, the concentration of PMS in the medium is usually about 0.1 mM.[30,37,73] Incubation with PMS is carried out at darkness, anaerobically, and/or in the presence of KCN to prevent oxidation of $PMSH_2$ through cytochrome oxidase.

Dimethyl Sulfoxide (DMSO)

Inclusion of dimethylsulfoxide (DMSO) into the incubation medium frequently facilitates the cytochemical demonstration of certain oxidoreductases, as was shown, e.g., with SDH,[74] NADH-, NADPH-, and lactate dehydrogenases, or NAD-linked MDH in animal tissues.[3] However, it inhibited NADP-linked MDH and both the NAD- and NADP-linked glutamate dehydrogenases.[3] The effect of DMSO is often concentration dependent: 5 to 10% of it enhanced, 15% inhibited SDH and NAD-linked isocitrate dehydrogenase staining reactions in the hyphae of a fungus.[75] According to Hanker,[3] many of the effects of DMSO can be explained by its molecular structure. The highly polarized DMSO molecules (with a relatively low dielectric constant) intercalating between associated water molecules render them more ordered, facilitating, thereby, the electron transfer from NAD(P)H and/or from the thiol groups at the active site of several dehydrogenases to an artificial acceptor. Changes in enzyme conformation as a consequence of the replacement by DMSO of water bound to protein might, however, lead to the inhibition of certain enzymes. Other effects of DMSO, such as acceleration of the penetration of the substrates into the cells,[74] are thought to be independent of the electron transfer facilitation. In plant ultracytochemistry, with the exception of monoamine oxidase (MAO) demonstration,[76] neither penetration of the media,[41,77] precision of localization, nor enzyme reactivity were improved with the use of DMSO.[41]

PREINCUBATION PROCEDURES

Incubation of Fresh Material Vs. Prefixation

Fixation, while preserving ultrastructure, unfortunately most often inactivates the dehydrogenases,[8,11,24] probably as a consequence of the modification of the thiol or other groups at the substrate-binding site of the enzyme or due to conformational changes.[3] The use of fresh plant material was considered by Jensen[8] as an absolutely essential prerequisite for successful demonstration of dehydrogenase activities, and the majority of light microscopic investigations concerning dehydrogenases with the use of tetrazoles were performed on chemically unfixed tissues.[26,30-35,73,78] The use of intact tissues for ultracytochemistry turned out to be problematic due to the conspicuous damage inflicted on the ultrastructure during incubation. On the other hand, even the mildest pretreatment of the samples with formalin eliminated or severely reduced enzyme activity.[26] The lack of a suitable electron acceptor and the seemingly irresolvable problem — either ultrastructure or enzyme activity — have caused a significant time-lag in plant dehydrogenase ultracytochemistry. Earlier findings, however, of the specific precipitate distribution in unfixed plant tissues, using ferricyanide as electron acceptor,[12] were later confirmed with the use of a more suitable method and even more so by prefixation of the material with 0.1% or less glutaraldehyde for 20 to 40 min.[13-15] Glutaraldehyde in higher concentrations, nevertheless, eliminated the reaction as was observed by other authors as well.[16] Only certain aldehydes (e.g., formaldehyde, glutaraldehyde) can be used for prefixation.[3,4,11,22,24]

The preserved activity of the aldehyde-fixed enzyme is reciprocal to the amount of cross-linkages created by the fixative, i.e., to the reactivity of the fixative (glutaraldehyde » formaldehyde).[22] The number of cross-linkages introduced by a given fixative is influenced, apart from its concentration, by many factors, as pH, temperature and duration of the fixation, etc. The speed of this process is in direct proportion to the rates of diffusion of the fixatives into the tissues (formaldehyde > glutaraldehyde). On the other hand, there are marked differences in the sensitivity of the individual enzymes to the effect of the fixatives. In animal tissues the fixation with formalin was best endured by the NAD(P)H-, lactate-, and malate dehydrogenases, while the NADP-linked dehydrogenases had the least tolerance to it, and NADP-tetrazolium reductase was found to be more resistant than NADH-tetrazolium reductase.[4] SDH is very sensitive to fixation, surviving only a very short (5 to 15 min) treatment in cold 0.7 to 2% formaldehyde.[4,69] Plant dehydrogenases are considered to be extremely sensitive to fixatives.[8,23,24] Nevertheless, SDH (together with MDH) and a plastid envelope-localized (NADH) ferricyanide reductase retained sufficient activities after a prefixation with 0.1% or less glutaraldehyde for 20 to 40 min.[13-15] Glycolate dehydrogenase activity in the mitochondria was not inhibited with 4% formaldehyde, but 1% glutaraldehyde wholly destroyed it,[17] and plastid MDH tolerated the effect of a short pretreatment with paraformaldehyde.[19] Although the major proportion of the inactivation occurs usually in the first few minutes of the fixation,[79] selectivity of the permeability barriers are not lost abruptly.[24] The cells and their compartments still might behave osmotically with many processes going on and/or upset for a shorter or longer period.[80,81] This happens particularly when using low fixative concentrations.

In order to avoid the problem of possible enzyme inactivation, fresh intact tissues are incubated frequently prior to any fixation. This approach will, however, raise certain problems. Permeability barriers will be left more or less intact and the diffusion rate of the ingredients of the medium through them might be strikingly different. As a consequence of the components' not arriving in the same time, in the required amounts and proportions at a given intracellular enzymatic site, aspecific precipitation and/or lack of reaction may occur. Deterioration of the fine structure seems to be almost inevitable and only a rather narrow surface layer will be well penetrated in both animal (30 to 200 μm)[47,82] and plant materials (root tissues three to four cell layers,[15] leaf tissues two to three cell layers[13,14]).

Although it is questionable[24] whether soluble enzymes and coenzymes are retained better[82] due to the integrity of plasma membranes, the retention of endogenous substrates by unfixed samples often results in specific reaction in the controls, incubated without exogenous substrate.[12,13,35,82] Nevertheless, this approach seems to be applicable for dehydrogenases bound to membranes and/or enclosed between the two limiting membranes, such as in mitochondria and plastids.[12-17] Neither incubation for prolonged times nor at elevated temperature (37°C) could, however, improve weak permeability.[12] Only pretreatment with a very low concentration of glutaraldehyde increased significantly the depth of the well-penetrated zone in lupine roots up to five to seven cell layers and, although not consistently, to some extent in leaf tissues as well.[13-15] Nevertheless, employing preincubation one should not forget that "unfortunately it is not possible to be sure whether enzymes in some location are more vulnerable to inactivation (by fixatives) than others and hence affect the overall picture."[79] Moreover, if the concentration of the fixative is lowered too much, enzymes might remain diffusible with the harmful consequences of localization artifacts.

Freezing

Another attempt to improve cell permeability apart from prefixation, DMSO, or mild sonication[17] is the freezing of the samples at about −70 to −60°C and cutting them with a cryostat at −25 to 15°C and into 6- to 15-μm-thick sections. The light microscopic demonstration of a whole range of enzymes[30,36,78] invalidates the earlier statement that frozen tissue enzymes are inactivated.[8] To avoid tissue damages through ice formation, pretreating the tissue with 5% polyvinylalcohol (PVA) for 1 to 2 hr at room temperature was recommended.[36] Addition (22% w/v) of this colloid tissue stabilizer into the incubating medium could not prevent the loss of formazan,[83,84] indicating that both the qualitative evaluation of the reaction and the measurement of enzyme activity should be made immediately at the time of incubation.[85] Moreover, freezing and thawing may activate latent enzymes or may increase enzyme activity resulting in activities different from those in vivo.[3,86] The loss of dehydrogenase activities that might occur with freezing was ascribed by Steponkus,[87] not to the dehydrogenase denaturation (SDH cooled at −70°C retained activity), but to cofactor or substrate limitations brought about, perhaps, by physiological or metabolic disruption within the cell.

The applicability of the ultracryotomy for unfixed tissues is limited by the presence of soluble enzymes and cofactors. Besides, freezing of unfixed tissues and particularly their inevitable subsequent thawing may cause unacceptable damage of the fine structure. Efforts to overcome the above-mentioned difficulties using, apart from PVA, polyvinylpyrrolidone, gelatine, agarose, semipermeable membranes, etc. yielded only partial results as yet.[4,84]

To sum up, fresh material is recommended to start with for cytochemical incubation, and only after demonstrating the presence of dehydrogenase under study in such samples to make attempts at improving fine structural preservation, starting with the use of very low concentration (0.1% of less) aldehyde in phosphate-buffered sucrose) or other osmoticum). To lessen structural damage, every preincubation is recommended to be carried out in the presence of phosphate-buffered sucrose.

POSTINCUBATION PROCEDURES

The postincubation wash should be accomplished with phosphate-buffered sucrose (two changes, 5 to 10 min each) for minimalizing structural damage and diffusion artifacts in the osmotically still active tissue samples. For the same reason it is recommended to carry out postfixation in 1 to 2% OsO_4, or 3 to 5% glutaraldehyde followed by 1 to 2% OsO_4, with fixatives dissolved in phosphate-buffered sucrose.

LIGHT MICROSCOPIC METHODS FOR DEMONSTRATION OF DEHYDROGENASES

LM/1: Succinate Dehydrogenase

The original method of Nachlas et al.,[88] using NBT for the demonstration of SDH activity in rat heart, has been employed, occasionally with modifications, for plant materials.

LM/1.1: The Method of Jensen[8] after Nachlas et al.[88]

Medium — This includes 10 mℓ of a stock solution of equal volumes of 0.2 M phosphate buffer (pH 7.6) and 0.2 M Na-succinate, added to 10 mℓ of an aqueous solution containing 10 mg of NBT.

Procedure — Fresh tissue pieces are incubated for 5 to 30 min at 37°C, washed for 1 min, and dehydrated in ethanol series starting with 30%. Subsequently, they are placed into xylol for 10 to 30 min, then mounted in balsam.

Results — The sites of activity (mitochondria) will appear deep blue.

Controls —

1. Incubation without substrate
2. Incubation in a medium with α-ketoglutarate, fumarate, or malate as substitutes for succinate
3. Incubation in a medium with succinate plus the inhibitor p-phenylene diamine (1 mM)
4. Incubation of a heat-inactivated tissue (heat inactivation at 80°C for 1 hr[88] or in boiling water for 1 min[31]): no reaction

Controls in addition —

5. Incubation in a medium also containing Na-malonate (8 mM);[88] pretreatment with malonate prior to incubation in a medium with malonate:[31] partial inhibition
6. Preincubation with 10 mM iodoacetate for 30 min:[88] total inhibition
7. Incubation in a medium also containing p-chloromercuribenzoate (PCMB) 0.1 mM,[32] 0.5 mM:[30] inhibition

LM/1.2: The Method of Sauter and Braun[32]

Medium — This includes 0.2 mℓ Na-succinate (1 M), 0.25 mℓ phosphate buffer (1/15 M, pH 7.0), 0.25 mℓ NBT (2 mg/mℓ distilled water), and 0.3 mℓ distilled water.

Procedure — Frozen sections (35 μm) from *Larix* stem were incubated for 10, 30, or 60 min at 37°C, washed briefly with buffer, and fixed in 4% formaldehyde for at least 30 min. After washing with buffer the samples were embedded into glycerol-gelatine.

Controls — This included omission of the substrate addition of PCMB (0.1 mM).

LM/1.3: The Method of Fosket and Miksche,[89] as Used by Chauhan and Lal[33]

Medium — This includes 0.05 M Na-succinate, 0.05 M phosphate buffer (pH 7.6), and 0.5 mg/mℓ NBT.

Procedure — Fresh free-hand sections from the shoot apical meristem of *Pinus lambertiana* were incubated in the above medium for 30 min (intact embryos for 2 to 8 hr) and finally fixed in 10% formaldehyde and mounted in glycerin jelly.

Control — This included incubation in the described solution containing 0.1 mM p-phenylene diamine.[89]

LM/1.4: The Method of Berlyn and Miksche[23]

Medium — This includes equal volumes of Na-phosphate-buffered 0.2 M Na-succinate and NBT solutions (10 mg/10 mℓ distilled water). Buffer: 12 mℓ from NaH$_2$PO$_4$ · H$_2$O (2.7 g to 100 mℓ distilled water) is added to 88 mℓ from Na$_2$HPO$_4$ · 7H$_2$O (5.36 g to 100 mℓ distilled water). Mix equal volumes of this phosphate buffer and Na-succinate solution (5.4 g in 100 mℓ distilled water).

Procedure — Cryostat sections (at $-20°$C at 8 to 15 μm) from fresh frozen material (at $-70°$C) are incubated for 30 to 60 min, rinsed in distilled water for 30 sec, and fixed in neutral 10% formalin, pH 7.0 (adjusted with 0.1 N NaOH). After dehydration through graded ethanols (30, 50, 70, 90, 100% for 5 min each) and two xylene baths, mounting is in Permount or equivalent mounting mixture.

Controls — These include heat inactivation; omission of the substrate from the medium (minus substrate control); inhibitor.

LM/1.5: The Method of Lehmann[30]

Medium — This included K-Na phosphate buffer (pH 7.2), 100 mM; EDTA, 2 mM; Na-succinate, 50 mM; KCN, 1.2 mM; NBT, 1.5 mM; and PMS, 0.1 mM.

Procedure — Frozen sections (16 to 26 μm) from the stem of *Cucurbita pepo* were incubated for 60 min at 37°C and examined immediately.

Controls — This includes omission of the substrate; addition of PCMB (0.5 mM) into a complete medium.

LM/2: Glyceraldehyde-3-Phosphate Dehydrogenase (Lehmann[30])

The medium contained triethanol amine-HCl-buffer (TRAB), pH 7.6, 50 mM; EDTA, 5 mM; NaHAsO$_4$, 2 mM; NAD, 1.8 mM; glyceraldehyde-3-phosphate, 3 mM; NBT, 1.5 mM; PMS, 0.1 mM; and glutathion, 1.2 mM. Incubation was for 10 min. For procedure and controls see ''LM/1.5''.

LM/3: Alcohol Dehydrogenase (Lehmann[30])

The medium contained Na-pyrophosphate, 50 mM; semicarbazid, 35 mM; glyine, 15 mM; ethanol, 10 mM; NAD, 1.8 mM; NBT, 1.5 mM; and PMS, 0.1 mM, pH 8.7. Incubation was for 10 min. For procedure and controls see ''LM/1.5''.

LM/4: Glucose-6-Phosphate Dehydrogenase (Lehmann[30])

The medium contained TRAB (pH 7.6), 50 mM; EDTA, 5 mM; glucose-6-phosphate, 15 mM; NADP, 1.4 mM; NBT, 1.5 mM; and PMS, 0.1 mM. Incubation was for 15 min. For procedure and controls see ''LM/1.5''.

LM/5: NADP-Linked Isocitrate Dehydrogenase (Lehmann[30])

The medium contained TRAB (pH 7.6), 50 mM; MgSO$_4$, 8 mM; D,L-isocitrate, 16 mM; NADP, 1.4 mM; NBT, 1.5 mM; and PMS, 0.1 mM. Incubation was for 25 min. For procedure and controls see ''LM/1.5''.

LM/6: NAD-Linked Isocitrate Dehydrogenase (Lehmann[30])

The medium contained TRAB (pH 7.6), 50 mM; EDTA, 5 mM; MgSO$_4$, 8 mM; D,L-isocitrate, 16 mM; glutathion, 4.8 mM; NAD, 1.8 mM; ADP, 4 mM; NBT, 1.5 mM; and PMS, 0.1 mM. Incubation was for 30 min. For procedure and controls see "LM/1.5".

LM/7: Malate Dehydrogenase (Lehmann[30])

The medium contained TRAB (pH 7.6), 50 mM; EDTA, 5 mM; D,L-malate, 60 mM; NAD, 1.8 mM; NBT, 1.5 mM; and PMS, 0.1 mM. Incubation was for 10 min. For procedure and controls see "LM/1.5".

LM/8: The Media Used by Sauter and Braun[32] with Stem of *Larix*

Medium — This includes 0.2 mℓ substrate solution (1 M), 0.25 mℓ phosphate buffer (1/15 M, pH 7.0), 0.25 mℓ NBT (2 mg/mℓ distilled water), 5 mg NAD, and 0.3 mℓ distilled water.

Substrates — These include DL-isocitrate (Na$_3$-salt)·2H$_2$O; L-malic acid; DL-glycerophosphate (Na$_2$-salt); ethanol (1 M); and DL lactate (Na salt). For procedure see "LM/1.2".

Controls — These include omission of the substrate from the medium; incubation without NAD or NADP; and addition of PCMB (0.1 mM) into the medium.

LM/9: Glutamate Dehydrogenase (Elhiti et al.[90])

Medium — The medium (1 mℓ) contained: L-glutamate (Na-salt), 42 mg; NAD, 0.5 mg, or NADP (sodium salt), 1.0 mg; NBT, 0.25 mg; 0.2 M glycine-NaOH buffer (0.25 mℓ), pH 9.6 for NAD-GDH, pH 9.0 for NADP-GDH, and distilled water, 0.75 mℓ.

Procedure — Small cubes (5 mm length) cut from fungus *Coprinus cinereus* were put into watered agar (1.5% w/v), frozen with liquid nitrogen, and sectioned into 20-μm-thick sections at −15°C. The mounted sections were incubated for 1 hr at 37°C, and after a washing with distilled water were mounted in 30% (v/v) glycerol.

LM/10: Lillie's Milieu for a Demonstration of Dehydrogenase Activities,[9] and Used by Catesson et al.[25] with *Dianthus* Bundles for NAD-Linked Malate Dehydrogenase

The medium used consisted of specific substrate (0.1 M), 15 mℓ; cofactor (0.1 M), 0.3 mℓ; 0.1 M Sörensen's buffer (pH 6.8 to 7.0), 9.0 mℓ; NBT or TNBT, 5 mg; and distilled water to make 30 mℓ.

LM/11: NADH- and NADPH-Diaphorase (Sauter and Braun[32])

The medium used by Sauter and Braun[32] with stem of *Larix* consisted of: NADH or NADPH, 1 mg; NBT (2 mg/mℓ distilled water), 0.25 mℓ; phosphate buffer (1/15 M, pH 7.0), 0.25 mℓ; and distilled water, 0.5 mℓ. For procedure and controls see "LM/1.2".

ELECTRON MICROSCOPIC METHODS FOR DEMONSTRATION OF DEHYDROGENASES

EM/1: NADH-Tetrazolium Oxidoreductase (Avers and Tkal[26])

Medium — This consisted of equal parts of 0.05% NBT, 0.5% NADH, 0.1 M phosphate buffer of pH 7.3, and distilled water.

Procedure — Primary root pieces (about 5 mm long) of *Phleum pratense* were incubated in the above medium for 5 to 12 min at 37°C, then fixed in 5% KMnO$_4$ for 2 min at room temperature. After dehydration with ethanol the samples were embedded in epoxide.

Results — Formazan deposits were principally restricted to the mitochondrial membranes, despite the ultrastructural damages. About half of the mitochondrial population of a cell was considered to be active for NADH- tetrazolium oxidoreductase.

EM/2: Succinate Dehydrogenase

The medium of Kerpel-Fronius and Hajós[47] elaborated for animal tissues, were used by Ekés,[12,13] Valanne,[16] and Bell,[18] for plant material. Note that better fine structural preservation and greater depth of penetration were achieved with the use of modified medium and procedure (see "Succinate Dehydrogenase and [Mitochondrial] Malate Dehydrogenase Activities..." that follows).

Medium — The medium (5 mℓ) mixed immediately before use, in the order listed, consisted of:

Stock solution	(mℓ)	Final conc (mM)
0.5 M K-Na tartrate (dissolved in the phosphate buffer used)	3.0	300
0.3 M CuSO$_4$ (in distilled water)	0.35	21
0.1 M Sörensen's phosphate buffer (pH 7.6)	0.8	76 (Together with the buffer of the tartrate solution)
1.0 M Na-succinate (in distilled water)	0.7	140
0.05 M K-ferricyanide (in distilled water)	0.15	1.5

Procedures

Roots of intact 6- to 7-day-old seedlings of *Lupinus luteus* L. were cut under a droplet of the incubation medium into segments of 1 mm length, beginning from the tips. After incubation for 20 to 30 min at room temperature, the segments were washed in 0.1 M phosphate buffer (pH 7.6) for 15 min and fixed with cold 5% glutaraldehyde in 0.1 M phosphate buffer, pH 7.2, for 2 to 2.5 hr. After a short washing in phosphate buffer and postfixation for 1 to 1.5 hr in 1% OsO$_4$ buffered with Millonig buffer at pH 7.4, they were dehydrated through graded ethanol series and propylene oxide and embedded in Durcupan.

Results — These include specific precipitation in the cristae and the space between the outer and inner limiting membranes of the mitochondria, and deposit in the periplastidal space (see EM/3). Observations were made on unstained sections or poststained for 1 min in uranyl acetate.

Controls —

1. Incubation for 20 min in the normal medium containing 140 mM Na-malonate diminished the amount of the precipitate.
2. Pretreatment of the blocks with 5 mM PCMB in 0.1 M phosphate buffer, pH 7.6, prior to the incubation in normal medium completely eliminated the reaction.
3. Omission of the substrate from the medium diminished the intensity of the reaction.

EM/2.2: Leaves
EM/2.2.1

White areas of mature leaves of a variegated mutant of *Betula pubescens* L.[16] were incubated, partially under vacuum, in the medium for 60 min at room temperature and processed similarly as described above. A somewhat better structural preservation was achieved with 1% sucrose added to the medium. Precipitation was seen in both the intracristal spaces and the outer compartment of the mitochondria. Occasionally, the plastid envelope also showed activity.

Control — Incubation in a complete medium containing 140 mM Na-malonate eliminated the reaction as did prefixation with 1.5% glutaraldehyde.

EM/2.2.2

The above medium was used for etiolated leaves of *Hordeum vulgare L.*,[13] but they were treated partially as described in EM/3.

EM/2.3: Fern Egg Cells[18]

Thin slices (about 0.5 mm thick) cut vertically from the archegoniate region of cordate gametophytes of *Pteridium aquilinum* were incubated for 1 hr at 35°C in the dark. After a brief washing in 0.05 M phosphate buffer (pH 7.6) at room temperature, the samples were drained, then fixed with 1% OsO_4 in the same buffer for 1 hr at 4°C, dehydrated in acetone, and embedded in Durcupan. Slices (about 1 mm thick), cut as described above, from *Dryopteris filix-mas* were pretreated in sterile tap water for 1 hr at 35°C with constant aeration in full daylight. Incubation and subsequent treatment were as for *Pteridium*.

Results — These included electron-dense reaction product in association with mitochondria. In *Pteridium*, envelopes of the nuclear evaginations and extra membrane around the egg occasionally contained precipitate. In *Dryopteris* a deposit was present in the plastid envelopes.

Controls — For *Pteridium*, after omission of the substrate from the medium or preincubation of the samples in 0.1 M Na-malonate (in phosphate buffer, pH 7.6) for 30 min at room temperature, before incubation in the normal medium containing Na-malonate (140 mM), only the extra membranes showed precipitate. For *Dryopteris*, minus-substrate control, and incubation in a normal medium with 14 mM succinate and 140 mM malonate resulted in a lack of the mitochondrial reaction, but did not affect the deposition in the plastid envelopes.

EM/3: Succinate Dehydrogenase and (Mitochondrial) Malate Dehydrogenase Activities, and Succinate- and Malate-Driven Ferricyanide Reductase Activity in the Plastid Envelope Compartment (Based on Experiments with Etiolated Leaves, Green Leaves, and Roots; Ekés[13-15])

Buffers

Three buffers included the following:

1. 0.1 M K-Na phosphate buffer, pH 7.6
2. 0.1 M K-Na phosphate-buffered (pH 7.6) 200 mM sucrose solution
3. 0.1 M K-Na phosphate-buffered (pH 7.6) 100 mM sucrose solution

Buffers are made with glass distilled water, and buffers 2 and 3 are made on the day of their use.

Medium — Basic incubation medium (10 mℓ) includes:

Stock solution	(mℓ)	Final conc (mM)
0.5 M K-Na tartrate in buffer 1	3.0	150
0.3 M $CuSO_4$ in glass distilled water	0.5	15
0.1 M buffer 1	4.8	78 (Together with buffer introduced with the tartrate)
1.0 M substrate Na-succinate, in glass distilled water (basic medium 1) Na-D,L-malate (basic medium 2)	1.4	140
NAD, if added	10 mg	1 mg/mℓ
0.05 M K-ferricyanide in glass distilled water	0.3	1.5

Individual stable stock solutions of phosphate buffer, tartrate, copper sulfate, and ferricyanide might be prepared previously and stored in refrigerator. The medium is prepared immediately before use with intense mixing of the ingredients in the given order. The solution obtained is of green color and clear. It is stable for several hours. The final pH of it is 6.9 to 7.0.

D,L-Malate (Na-salt) solution (1 M) is prepared from D,L-malic acid (mw 134.09) neutralized with NaOH and buffer 1. The final pH is adjusted to 7.0.

Fixatives

These include 5% glutaraldehyde (cleaned by vacuum distillation) in buffer 2; and 2% OsO_4 in buffer 2 or 3.

Procedures

Squares (*circa* 1 mm^2 pieces) are excised from leaf material and segments (*circa* 0.2 to 1 mm length) are cut transversally and/or longitudinally from roots, beginning with the tips, under a droplet of the solution they are to be treated with first.

Standard Procedure 1: Without any Prefixation of the Samples

- Standard incubation: performed for a period of 20 to 40 min at room temperature (eventually under a mild vacuum)
- Washing: in two changes of cold buffer 2 for 10 min each
- Postfixation: in chilled 2% OsO_4 (in buffer 2 or 3) for at least 1 hr at +4°C; alternatively, might be preceded by a fixation with cold 5% glutaraldehyde (in buffer 2) for 1 to 2 hr at +4°C, followed by rinsing twice in chilled buffer 2 for 5 to 10 min each
- Washing in buffer 2 or 3 for 10 to 15 min at room temperature
- Dehydration: in a graded ethanol series for 10 min each, and propylene oxide two changes for 20 min each
- Embedding: in Durcupan

Ultrathin sections were examined unstained or stained with uranyl acetate.

Standard Procedure 2: With a Prefixation of the Samples

Prefixation of the samples with cold 0.1% (or less concentrated) glutaraldehyde in buffer 2 for 20 to 40 min at 4°C. After washing with two changes of cold buffer 2 for 10 to 15 min each, subsequent treatment was as standard incubation and processing.

Results — Only the mitochondria and the plastids showed staining in the cytochemically evaluable parts of the samples (Figures 1 and 2). Electron-dense precipitate due to the activities of the malate dehydrogenase (basic medium 2) and succinate and malate dehydrogenase (basic medium 1) was confined specifically to the intracristal space and the outer compartment of the mitochondria. Deposition in the plastid envelope area resulted from the activity of an envelope-localized malate- and succinate-driven ferricyanide reductase, supposedly NADH-ferricyanide oxidoreductase.

Control experiments (Note: "Remarks on the Control Experiments" to follow)
1. and 2. Controls for the Specificity of the Reaction

1. Incubation without the electron acceptor (ferricyanide is replaced by equal volume of distilled water in the basic medium) for 20 to 40 min at room temperature. Subsequently treated as for standard procedures.
2. Product and capturing agent precipitation control: incubation with tartrate-chelated

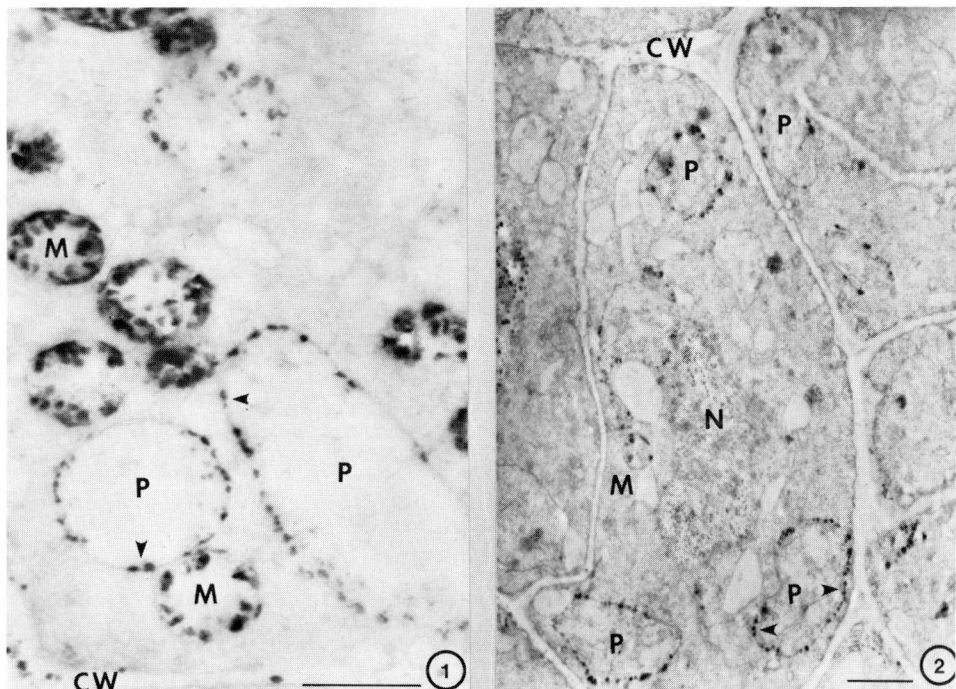

FIGURES 1. and 2. Distribution of reaction product within root (Figure 1) and leaf (Figure 2) cells of *Lupinus albus* incubated unfixed for 30 min in a succinate- (Figure 1) or malate-containing (Figure 2) basic medium by the method of Ekés (see EM/3). Enzymatically produced deposit of copper ferrocyanide is confined to the plastid (P) envelope area (arrows), and mitochondria (M). Note the uneven intensity of staining among the mitochondria of the root cell (Figure 1). N = nucleus; CW = cell wall.

copper in the used concentrations (150 and 15 mM, respectively, in buffer 2) for 10 min, and after a short rinse in buffer 2, with 5 mm ferrocyanide (in buffer 2) for 10 min.

3. Controls to Elucidate the Enzymic Nature of Precipitation

3.1. Heating the samples in buffer 2 for 5 min at 50 to 80°C. Subsequently treated as for standard procedures.
3.2. Prefixation with 1.25, 0.5, 0.1, 0.05, and 0.025% (distilled) glutaraldehyde in buffer 2 for 20 to 40 min at +4°C. After rinsing with buffer 2 twice for 15 min subsequently treated as for standard procedures.
3.3. Incubation for prolonged time.
3.4. Incubation at elevated temperature (37°C).
3.5. Standard procedures with lower and higher amount of substrates.
3.6. Incubation with omission of the substrate from the medium (substrate is replaced by equal volume of distilled water). Subsequent treatment as for standard procedures.
3.7. Preincubation of the samples with 3 mM ferricyanide (in buffer 2) for 25 min at room temperature. After washing with buffer 2 (two changes, for 10 min each) incubated:
3.7.1. In a substrate-free medium (3.6.).
3.7.2 In succinate-containing basic medium 1.
3.7.3. In succinate- and NAD-containing basic medium 1.

(It is advisable to run similar experiments with malate-containing basic medium 2 as well.)

4. and 5. Inhibitor Controls

4.1. Preincubation of the samples with 100 mM Na-malonate for 20 min, followed by incubation in basic media in which part of the buffer 1 is replaced by 140 mM Na-malonate, for 40 min at room temperature. Subsequently treated as for standard procedures.

4.2. Controls with sulfhydryl reagents.

4.2.1. Pretreatment with N-ethylmaleimide (NEM) at a concentration of 49 mg/10 mℓ of buffer 2 for 30 min at 37°C. After washing with buffer 2 for 10 min incubated and processed as for standard procedures.

4.2.2. Pretreatment with p-chloromercuribenzoate (PCMB) (5, 2, 1, and 0.1 mM) dissolved in buffer 2 for 15 min at room temperature. After rinsing with two changes of buffer 2 (10 min each) incubation and processing as for standard procedures. PCMB solution should be prepared dissolving it by a small amount of NaOH then carefully lowering the pH with the addition of HCl drop by drop just to the appearance of a very weak precipitation. The needed concentration of PCMB is achieved by the addition of buffer 2.

5. Preincubation of the samples with 10, 5, 1, and 0.5 mM KCN (neutralized, in buffer 2) for 15 min at room temperature. After washing with two changes of buffer 2 for 10 to 15 min each, treated and processed as for standard procedures.

EM/4: Malate Dehydrogenase

Santos and Salema,[19] using the method of Wenzel and Behrisch,[91] elaborated for electron microscopic investigation of oxidoreductases in rat heart muscle.

Medium — The medium (5 mℓ) was prepared immediately prior to use:

Stock solution	mℓ	Final concentration (mM)
0.1 Sörensen's phosphate buffer (pH 7.6)	1.0	80 (together with the buffer of the tartrate solution)
0.5 M K-Na tartrate	3.0	300
0.3 M $CuSO_4$	0.35	21
0.1 M D,L-malate	0.5	10
NAD or NADP		1—5 mg/5 mℓ of medium
0.05 M K-ferricyanide	0.15	1.5

The pH of the medium is adjusted to pH 7.0.

Procedure

Sedum telephium leaf samples were shortly fixed with (para)formaldehyde, before incubation for 2 hr at 3°C.

Results — Specific electron-dense precipitation was observed in the microtubular inclusion of the chloroplasts with either NAD or NADP-containing medium. It was absent if the tissue was not previously fixed.

Controls

Heating the samples at 70 to 80°C during formaldehyde prefixation, or incubation in a medium without substrate resulted in a lack of reaction.

EM/5: Glycolate Dehydrogenase (Breezley et al.[17])

Medium — The incubation medium (10 mℓ) was prepared immediately before use in the order listed:

Stock solution	mℓ	Final concentration (mM)
0.2 M K-phosphate (pH 7.2)	2.5	90 (together with that of the tartrate solution)
25 mM K-Na tartrate	2.0	5
25 mM CuSO$_4$	2.0	5
50 mM K-ferricyanide	0.4	2
0.5 M Na-glycolate	0.4	20
Distilled water	2.7	

Procedure

Chlamydomonas reinhardii cells, grown in cultures, after harvesting were rinsed twice with 50 mM K-phosphate (pH 7.2) and resuspended in this buffer (1 mℓ packed cells/25 mℓ final volume). After mild disruption of the cells with sonication for 5 sec they were incubated in the dark at 25°C for 20 min, in other cases for 2 or 5 min rinsed with 50 mM K-phosphate (pH 7.2), embedded in 2% agar and rinsed two more times. Fixation of the samples with 2% glutaraldehyde (in 50 mM K-phosphate buffer, pH 7.2) for 1 hr at room temperature was followed by washing for at least 60 min with four or five changes of 50 mM K-phospate (pH 7.2) and postfixation with 2% OsO$_4$ buffered with 50 mM K-phosphate (pH 7.2) for 60 min at room temperature. After dehydration by acetone the samples were embedded in Spurr's low viscosity resin, and silver gray sections were counterstained with 2% aqueous uranyl acetate for 10 min, and lead citrate for 5 min.

Results

Specific deposit in the outer compartment of the mitochondria.

Controls —

1. Incubation with other potential substrates. Instead of glycolate 0.5 MD(−)-lactate: positive reaction; 0.5 M L(+)-lactate: almost no reaction.
2. Omission of substrate from the medium replacing it with 0.4 mℓ H$_2$O: no specific deposition.
3. Preincubation with 20 or 100 mM Na-oxamate dissolved in 50 mM K-phosphate (pH 7.2) inhibited the reaction. Prefixation with 4% formaldehyde for 10 to 30 min increased aspecific precipitation in the chloroplasts, 1% glutaraldehyde completely eliminated the specific deposit in mitochondria. Spectrophotometrical assay of cell-free preparations corroborated well with the cytochemical observations.

REMARKS ON THE CONTROL EXPERIMENTS

In order to verify the cytochemical findings experiments with alternative approaches for elucidating the localization of the same enzyme should be run simultaneously. The usual practice, if it is accomplished at all, to relate cytochemical results to biochemical data, is not free of serious interpretational uncertainties. Indeed, while cytochemistry deals with tissue or cell samples with enzymes localized *in situ,* every biochemical analysis involves the unavoidable step of physical disruption of the cells and/or organelles. Disruption, fractionation, and purification inevitably lead to structural damages, diffusion, changes in compartmentalization, and possibly to denaturation of enzymes, etc.[3,24] Isolated and purified enzymes in optimal conditions of pH and temperature, well supplied with substrates, coenzymes, etc., might behave in absolutely different ways than in their native milieu within the cell. Moreover, individual variations between different cell types and between similar organelles in the same cell might have been lost.[24] Much of the information from biochemical studies has been incorrectly interpreted owing to the unobservance of these circumstances. Consequently, the lack of coincidence with biochemical data should not always be considered as indicative of the inadequacy of the enzyme cytochemical approach. The use of intact tissue sections for biochemical studies,[92] as well as involving the methods of immunocytochemistry and autoradiography, might be promising for overcoming some of these difficulties. In the following sections those cytochemical controls are commented on that were performed in experiments for demonstration of malate- and succinate-driven ferricyanide reductase in plastid envelope compartment by Ekés[12-15] (see EM/2.1; EM/3). Nevertheless, it is hoped that these comments might be helpful for other cytochemical experiments as well. Numbers in the parentheses refer to the corresponding controls in experiments EM/3.

Controls to Prove that the Precipitate in the Mitochondria and the Plastid Envelope Compartment Resulted from a Specific Reaction

No precipitation occurred when incubation was carried out in a medium (control 1) lacking the electron acceptor (ferricyanide). This proves the absolute necessity of an electron acceptor for the formation of electron-dense deposit.

Incubation with chelated copper, followed by ferrocyanide (control 2) was performed in order to elucidate whether the primay product (ferrocyanide) or the capturing agent (Cu^{2+}) binds preferentially to certain structures. The patchy appearance of the precipitate scattered throughout the cells proves that an electron-dense product might be deposited anywhere, provided ferrocyanide and copper ions have an access to each other, without preference for any structure. Diffusion of the electron-dense product from reacted mitochondria into the plastids could be excluded; as (usually) no other cytomembranes were stained, copper ferrocyanide is insoluble under the conditions of the cytochemical assays, and nonreactive plastids could often be seen near strongly reacted mitochondria.

Controls to Elucidate the Enzymic Nature of the Precipitation

Heating the samples at 50 to 80°C (control 3.1) resulted in a lack of specific precipitation, indicating the enzymic nature of the deposits seen without heat pretreatment: enzymes are usually inactivated by elevated temperatures. Prefixation in 1.25%, 0.5% glutaraldehyde (control 3.2) resulted only in aspecific precipitation showing the extreme sensitivity of the reaction to glutaraldehyde fixation. This feature is considered to be characteristic of the majority of dehydrogenases,[3,4,22] especially those of plant origin.[8,16,17,23,24,26] Pretreatment with 0.1% (and less) glutaraldehyde while preserving better ultrastructure did not alter the overall staining pattern.[13-15]

Controls 3.3-3.7 were designed to check the dependence of the precipitate production upon the duration (control 3.3), the temperature of the incubation (control 3.4), and on the amount of the exogenous and endogenous substrates (controls 3.5., 3.6., and 3.7). The amount of the precipitate was increased, as a rule, by the extension of the incubation time (control 3.3), by elevated temperature of incubation (control 3.4), and often by higher concentrations of the exogenous substrates (control 3.5) in the medium. Omission of the substrate from the medium ("minus substrate control") (control 3.6) is the most frequently used method to show the substrate dependency of the enzyme in animal[2-4,11] and plant materials[24] as well. Quite often, however, reduction of NAD(P) and electron acceptors (tetrazolium salts, ferricyanide) occurs even in the absence of any added substrate due to the functioning of a so-called "nothing dehydrogenase".[2,11,22,24] Its activity increases with the rise of pH from 7.0 up to pH 9.0, and can be inhibited by freezing or by sulfhydryl reagents (at a concentration of 0.1 M). In many cases this activity might be caused by nonenzymatic reduction of electron acceptors due to the presence of protein-bound thiols, by glutathione, or through the active site of alcohol dehydrogenase. It is possible, however, that partly or wholly the precipitate arises from the enzymatic action of the large amount of yet unidentified endogenous tetrazolium (ferridyanide) reductases on endogenous substrates.

In the histochemistry of animal tissues "nothing dehydrogenase" is considered to be taken into account only when the pH of the medium is above 7.4 (especially if the incubation is prolonged) and in the presence of e.g., PMS.[21,51] Plant tissues, however, contain organic acids (substrates) in high concentrations[8] (10 to 100 times more as compared to those found in animals),[93] and due to it the endogenous substrate-driven "nothing dehydrogenase" activity might be quite considerable normally. As a consequence of it and depending on the physiological state of the sample, intense precipitation without added substrate might take place. Omission of the substrate from the medium sometimes only slightly diminishes the intensity of the reaction (control 3.6). The large quantity of endogenous substrates is considered as one of the most difficult tasks to overcome in plant cytochemistry.[35] Starvation of the samples in phosphate buffer for a longer period (about 20 hr) in darkness to exhaust all endogenous oxidizable substrates, did not lead to considerable deterioration of the fine structure of root tips, but resulted in unacceptable structural damages in leaf pieces. In order to eliminate endogenous substrates and other compounds capable to reduce ferricyanide, preincubation of the samples in ferricyanide is sometimes recommended (control 3.7).[24,40] The ferrocyanide produced must be washed out from the tissues before the proper incubation. In these experiments (control 3.7.1 to 3.7.3) many of the soluble or weakly bound enzymes might be washed out and obviously all amount of the oxidizable endogenous substrate(s) was consumed up and partially or wholly washed out during the rinsing procedure. As a result, no reaction could be seen after incubation in a substrate-free medium (control 3.7.1). The weak reactivity of the mitochondria in the presence of added succinate (control 3.7.2) and the stronger one when NAD was added to the succinate-containing medium (control 3.7.3) indicate that the bulk of the deposit in mitochondria arose by the activity of some (mitochondrial) NADH-ferricyanide reductase, presumably, via NADH generated from NAD by the action of malate dehydrogenase and/or NAD-malic enzyme and succinate dehydrogenase. Deposition in plastid envelope area was also promoted by the addition of NAD, supporting, thereby, the idea that the direct electron donor for this reaction is NADH.[13-15] In a control experiment carried out similarly with succinate as the substrate, only mitochondrial staining was occasionally seen in etiolated barley leaf.[13]

Inhibitor Controls

The vital importance of adequate inhibitor controls should never be underestimated. Existing methods have, however, serious limitations. As stated by Borgers,[94] it would be highly urgent "to carefully control specificity of the reaction by employing various candidate substrates in various concentrations (if possible the natural substrate[s]), and by using only specific control reactions. Very often the only control reaction exists in omitting the substrate which has only very limited value. Inhibitors of the enzymatic reaction should be employed, provided (1) they do not interact at all with one of the components of the medium; (2) they inhibit at least 90% of the biochemically detectable activity; (3) they are only inhibitory for the enzyme under investigation."

Controls with malonate (control 4.1) a competitive inhibitor of SDH, usually diminished or eliminated the deposit formation both in the mitochondria and plastids, as was observed by Valanne[16] as well. Plastid envelope staining remained, however, in the experiments of Bell[18] with *Dryopteris*. Diminished activity sometimes resulted in a slightly diffuse precipitation. It must be borne in mind, however, that the effect of malonate is complex in living

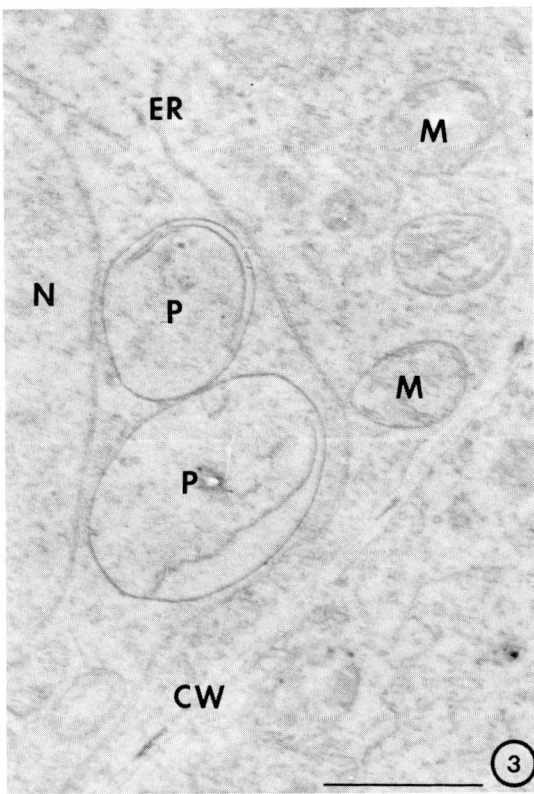

FIGURE 3. After pretreatment with 1 mM PCMB prior to incubation in the basic medium no electron-dense precipitate was observed (unfixed root; malate as substrate). P = plastid; M = mitochondrion; ER = endoplasmic reticulum; N = nucleus; CW = cell wall. See EM/3.

systems, depending on the concentration of the substrate, the metabolic state of the sample, etc.[59,95]

Sulfhydryl reagents N-ethylmaleimide (NEM)[13] (control 4.2.1) and p-chloromercuribenzoate (PCMB) (control 4.2.2; Figure 3) eliminated the deposit production. In conformity with literature data,[2,3,47,50] these reagents act by blocking –SH groups of the enzymes. The lack of reaction indicates the importance of –SH groups in functioning of the flavin prosthetic groups in the enzymes involved.

Pretreatments of the samples with KCN (control 5) often resulted in a rather damaged ultrastructure in the outer part of the penetrated zone containing cells with frequent aspecific precipitation. Usually, only the innermost penetrated cell layers contained reacted plastids and reacted mitochondria (Figure 4); outwards the amount of the reacted plastids decreased. Prefixation of the material with glutaraldehyde, although it depressed the intensity of the precipitation, did not alter its distribution pattern.

Owing to the manifold and complex effects of KCN[27,96,97] it is rather difficult to explain these observations. In fact, KCN is a very unspecific inhibitor, and in low concentrations it can inhibit, apart from cytochrome oxidase, a number of heme-, copper-, molybdenum-, and other metal-containing enzymes, inhibiting them by several different mechanisms. By removing the copper from the enzyme it inhibits ascorbate oxidase and may cease photosynthetic electron transport removing the copper from the plastocyanin. It may affect enzymes

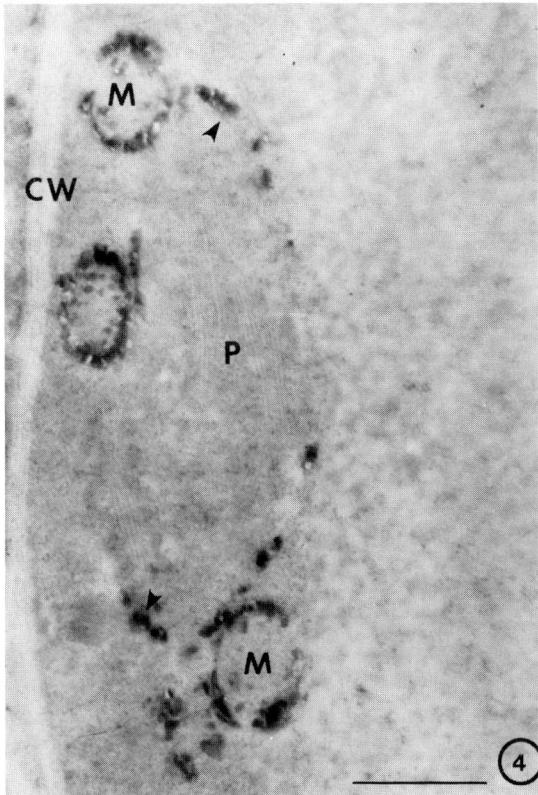

FIGURE 4. Strong reaction in the chloroplast envelope area (arrows) and in the mitochondria (M) after pretreatment with 10 mM KCN followed by incubation in a malate-containing basic medium (unfixed leaf). CW = cell wall. See EM/3.

which contain pyridoxal phosphate and involve Schiff-base intermediate (e.g., glutamate decarboxylase), and indirectly can inhibit ribulose diphosphate carboxylase/oxygenase, the key enzyme in the Calvin cycle of photosynthetic CO_2 fixation. The sensitivity of the assimilatory nitrate reductase to cyanide was found to be dependent on the oxidation state of the enzyme. Cyanide may shunt glucose to the pentose phosphate pathway while decreasing the rate of glycolysis and inhibiting the tricarboxylic acid cycle.[96] The overall effect of cyanide depends on its concentration and, also, on other factors including pH and the levels and distribution of detoxifying enzymes.[96]

It cannot be excluded that (control 5) the easily penetrating cyanide, acting as a reducing agent, broke some essential disulfide links in the enzyme(s) and/or progressively inactivated SDH[97] in the outer layers of the tissue samples during preincubation. Subsequent rinsing, in addition, removed many of the soluble substrates, coenzymes, and possibly enzymes. As a result of these treatments, the formation of ferrocyanide was blocked or extremely delayed and its slow accumulation was balanced by the rate of its diffusion into the surrounding medium. Since cyanide has better penetrability than ferricyanide,[63] the presence of mitochondrial and plastid envelope-localized precipitate in the inner cells proves that cyanide in adequate conditions does not affect ferricyanide-reductase activities in these organelles.

Summarizing the results of experiments, it is supposed that a flavoprotein oxidoreductase (presumably, NADH-ferricyanide oxidoreductase) is present in the plastid envelope compartment, either loosely attached to the membrane(s) or enclosed in the periplastidal space.[13-15] Either alone or as a part of a larger complex this enzyme might be involved in various diaphorase reactions of the plastids e.g., taking part in the iron metabolism of these organelles as a Fe/III/reductase and/or functioning as the diaphorase of the nitrate reductase. Although no complementary biochemical experiments have been carried out yet, these assumptions are in conformity with some biochemical data.[98-101]

REFERENCES

1. **Dixon, M., Webb, E. C., Thorne, C. J. R., and Tipton, K. F.,** *Enzymes,* 3rd ed., Longman Group, London, 1979.
2. **Totovič, V.,** Oxydoreductasen, in *Methoden für den elektronenmikroskopischcytochemischen Enzymnachweis. Methodensammlung der Elektronenmikroskopic,* Schimmel, G. and Vogell, W., Eds., Wissenschaftliche Verlagsgesellschaft MBH, Stuttgart, 1971—1973, 141.
3. **Hanker, J. S.,,.** Oxidoreductases, in *Electron Microscopy of Enzymes. Principles and Methods,* Vol. 4, Hayat, M. A., Ed., Van Nostrand Reinhold, New York, 1975.
4. **Lojda, Z., Gossrau, R., and Schiebler, T. H.,** *Enzyme Histochemistry. A Laboratory Manual,* Springer-Verlag, Berlin, 1979.
5. **Bisalputra, T., Broun, D. L., and Weier, T. E.,** Possible respiratory sites in a blue-green alga *Nostoc sphaericum* as demonstrated by potassium tellurite and tetranitro-blue tetrazolium reduction, *J. Ultrastruct. Res.,* 27, 182, 1969.
6. **Shnitka, T. K. and Talibi, G. G.,** Cytochemical localization of ferricyanide reduction of α-hydroxy acid oxidase activity in peroxisomes of rat kidney, *Histochemie,* 27, 137, 1971.
7. **Raap, A. K. and Van Duijn, P.,** Studies on the phenazine methosulphate-tetrazolium capture reaction in NAD(P)$^+$-dependent dehydrogenase cytochemistry. II. A novel hypothesis for the mode of action of PMS and a study of the properties of reduced PMS, *Histochem. J.,* 15, 881, 1983.
8. **Jensen, W. A.,** *Botanical Histochemistry. Principles and Practice,* Freeman, San Francisco, 1962, chap. 15.
9. **Gabe, M.,** *Histological Techniques,* Masson, Paris, 1976, 652.
10. **Berchtold, J.-P.,** Ultracytochemical demonstration and probable localization of 3-β-hydroxysteroid dehydrogenase activity with a ferricyanide technique, *Histochemistry,* 50, 175, 1977.
11. **Wohlrab, F., Seidler, E., and Kunze, K. D.,** *Histo- und Zytochemie Dehydrierender Enzyme. Grundlagen und Problematik,* Johann Ambrosius Barth, Leipzig, 1979.
12. **Ekés, M.,** Electron-microscopic-histochemical demonstration of succinic-dehydrogenase activity in root cells of yellow lupine, *Planta,* 94, 37, 1970.
13. **Ekés, M.,** Ultrastructural demonstration of ferricyanide reductase (diaphorase) activity in the envelopes of the plastids of etiolated barley (*Hordeum vulgare* L.) leaves, *Planta,* 151, 439, 1981.
14. **Ekés, M.,** Ultracytochemical demonstration of malate- and succinate-driven ferricyanide reductase (diaphorase) activity in the chloroplast envelope of young white lupine (*Lupinus albus* L.) leaves, in *Proc. 8th Eur. Congr. on Electron Microscopy,* Vol. 3, Csanády, Á., Röhlich, P., and Szabó, D., Eds., Programme Committee of the 8th Eur. Congr. on Electron Microscopy, Budapest, 1984, 2119.
15. **Ekés, M.,** Effects of glutaraldehyde fixation on mitochondrial- and plastid envelope-localized ferricyanide reductase activities in white lupine (*Lupinus albus* L.) root tips, in *Proc. 8th Eur. Congr. on Electron Microscopy,* Vol. 3, Csanády, Á., Röhlich, P., and Szabó, D., Eds., Programme Committee of the 8th Eur. Congr. on Electron Microscopy, Budapest, 1984, 2121.
16. **Valanne, N.,** Peripheral structures of plastids and ultrastructural localization of acid phosphatase and succinic dehydrogenase in a variegated *Betula pubescens* mutant, *Can. J. Bot.,* 53, 1072, 1975.
17. **Beezley, B. B., Gruber, P. J., and Frederick, S. E.,** Cytochemical localization of glycolate dehydrogenase in mitochondria of *Chlamydomonas, Plant. Physiol.,* 58, 315, 1976.
18. **Bell, P. R.,** Demonstration of succinic dehydrogenase in mitochondria of fern egg cells at electron microscope level, *Histochemistry,* 62, 85, 1979.
19. **Santos, I. and Salema, R.,** Cytochemical localization of malic dehydrogenase in chloroplasts of *Sedum telephium,* in *Electron Microscopy 1980,* Vol. 2, *Proc. 7th Eur. Congr. on Electron Microscopy,* Brederoo, P. and de Priester, P., Eds., 7th Eur. Congr. on Electron Microscopy Foundation, Leiden, 1980, 246.
20. **Kurkova, E. B. and Verkhovskaya, M. L.,** Redox components of plant cell plasmalemma, *Fiziol. Rastenij,* 31, 496, 1984.

21. **Pearse, A. G. E.,** *Histochemistry. Theoretical and Applied,* Vol. 2, 3rd ed., Churchill Livingstone, Edinburgh, 1972.
22. **Geyer, G.,** *Ultrahistochemie. Histochemische Arbeitsvorschriften für die Elektronenmikroskopie,* Zweite, überarbeitete und erweiterte Auflage, VEB Gustav Fischer, Verlag, 1973.
23. **Berlyn, G. P. and Miksche, J. P.,** *Botanical Microtechnique and Cytochemistry,* Iowa State University Press, Ames, 1976, 290.
24. **Sexton, R. and Hall, J. L.,** Enzyme cytochemistry, in *Electron Microscopy and Cytochemistry of Plant Cells,* Hall, J. L., Ed., Elsevier/North-Holland, Amsterdam, 1978, chap. 2.
25. **Catesson, A.-M., Czaninski, Y., and Monties, B.,** Caractères histochimiques des peroxydases pariétales dans les cellules en cours de lignification, *C. R. Acad. Sci. Ser. D,* 286, 1787, 1978.
26. **Avers, C. J. and Tkal, M. M.,** Intracellular mitochondrial variation in enzyme activity as shown by histochemical studies using light and electron microscopy, *J. Histochem. Cytochem.,* 11, 157, 1963.
27. **Seidler, E.,** Zum Mechanismus der Tetrazoliumsalzreduktion und Wirkungsweise des Phenazinmethosulfates, *Acta Histochem.,* 65, 209, 1979.
28. **Kalina, M. and Palmer, J. M.,** The reduction of tetrazolium salts by plant mitochondria, *Histochemie,* 14, 366, 1968.
29. **Ponti, V., Dianzani, M. U., Cheesman, K., and Slater, T. F.,** Studies on the reduction of nitroblue tetrazolium chloride mediated through the action of NADH and phenazine methosulphate, *Chemico-Biol. Interact.,* 23, 281, 1978.
30. **Lehmann, J.,** Zur Lokalisation von Dehydrogenasen des Energiestoffwechsels im Phloem von *Cucurbita pepo* L., *Planta,* 111, 187, 1973.
31. **Hébant, C., Guiraud, R., Barthonnet, J., and Ba, A.,** Le phloeme de *Lycopodium clavatum:* organisation, ultrastructure et histochimie, *Can. J. Bot.,* 56, 2973, 1978.
32. **Sauter, J. J. and Braun, H. J.,** Cytochemische Untersuchung der Atmungsaktivitat in den Strasburger-Zellen von *Larix* und ihre Bedeutung fur den Assimilattransport, *Z. Pflanzenphysiol.,* 66, 440, 1972.
33. **Chauhan, E. and Lal, M.,** Localization of some hydrolases and succinate dehydrogenase in the sporophyte-gametophyte junction in *Physcomitrium cyathicarpum* Mitt., *Ann. Bot.,* 50, 763, 1982.
34. **Hébant, C. and Guillaume, C.,** Transfer cells in sporophytic axes of *Equisetum arvense:* ultrastructure and histoenzymology, *Flora,* 174, 137, 1983.
35. **Sutcliffe, J. F. and Sexton, R.,** Enzymatic changes during the differentiation of tissues in young pea roots, in *Structure and Function of Primary Root Tissues. Proc. Symp. Tatranská Lomnica, Sept. 7—10, 1971, Czechoslovakia,* Kolek, J., Ed., Veda, Publishing House of the Slovak Academy of Sciences, Bratislava, Czechoslovakia, 1974, 203.
36. **Kalina, M.,** Reduction of tetrazolium salts catalyzed by soluble disphorase in rat liver and embryos of *Vicia faba, Histochemie,* 7, 8, 1966.
37. **Hadačová, V.,** Contribution to the estimation of malic dehydrogenase isoenzymes in the root growth zones of *Vicia faba* L., *Biol. Plant.,* 14, 186, 1972.
38. **Karnovsky, M. J. and Roots, L.,** A "direct-coloring" thiocholine method for cholinesterases, *J. Histochem. Cytochem.,* 12, 219, 1964.
39. **Hanker, J. S., Yates, P. E., Clapp, D. H., and Anderson, W. A.,** New methods for the demonstration of lysosomal hydrolases by the formation of osmium blacks, *Histochemie,* 30, 201, 1972.
40. **Burke, J. J. and Trelease, R. N.,** Cytochemical demonstration of malate synthase and glycolate oxidase in microbodies of cucumber cotyledons, *Plant Physiol.,* 56, 710, 1975.
41. **Trelease, R. N., Becker, W. M., and Burke, J. J.,** Cytochemical localization of malate synthase in glyoxysomes, *J. Cell. Biol.,* 60, 483, 1974.
42. **Higgins, J. A. and Barrnett, R. J.,** Cytochemical localization of transferase activities carnityl acetyltransferase, *J. Cell Sci.,* 6, 29, 1970.
43. **Higgins, J. A. and Barrnett, R. J.,** Fine structural localization of acetyltransferases the monoglyceride and α-glycerophosphate pathways in intestinal absorptive cells, *J. Cell Biol.,* 50, 102, 1971.
44. **Kirchanski, S.,** Copper ferricyanide localization of photosystem II in glutaraldehyde fixed and unfixed chloroplasts, *J. Ultrastruct. Res.,* 57, 113, 1976.
45. **Ogawa, K. and Saito, T.,** New copper ferrocyanide method for the ultracytochemical demonstration of the succinic dehydrogenase activity, *J. Histochem.,* 14, 750, 1966.
46. **Ogawa, K., Saito, T., and Mayahara, H.,** The site of ferricyanide reduction by reductases within mitochondria as studied by electron microscopy, *J. Histochem. Cytochem.,* 16, 49, 1968.
47. **Kerpel-Fronius, S. and Hajós, F.,** The use of ferricyanide for the light and electron microscopic demonstration of succinic dehydrogenase activity, *Histochemie,* 14, 343, 1968.
48. **Estabrook, R. W.,** Studies of oxidative phosphorylation with potassium ferricyanide as electron acceptor, *J. Biol. Chem.,* 236, 3051, 1961.
49. **Klingenberg, M.,** The ferricyanide method for elucidating the sidedness of membrane-bound dehydrogenases, in *Methods in Enzymology,* Vol. 56, Fleischer, S. and Packer, L., Eds., Academic Press, New York, 1979, 229.

50. **Kugler, P. and Wrobel, K.-H.** Studies on the optimalisation and standardisation of the light microscopical succinate dehydrogenase histochemistry, *Histochemistry,* 57, 47, 1978.
51. **Lukaszyk, A.,** A method for histochemical demonstration of α-glycerophosphate-ferricyanide oxidoreductase activity, *Folia Histochem. Cytochem.,* 9, 167, 1971.
52. **Löw, H. and Crane, F. L.,** Redox functions in plasma membranes, *Biochim. Biophys. Acta,* 515, 141, 1978.
53. **Lin, W.,** Further characterization on the transport property of plasmalemma NADH oxidation system in isolated corn root protoplasts, *Plant Physiol.,* 74, 219, 1984.
54. **Barr, R., Crane, F. L., and Craig, T. A.,** Transmembrane ferricyanide reduction in tobacco callus cells, *J. Plant Growth Regul.,* 2, 243, 1984.
55. **Crane, F. L., Roberts, H., Linnane, A. W., and Löw, H.,** Transmembrane ferricyanide reduction by cells of the yeast *Saccharomyces cerevisiae, J. Bioenerget. Biomembr.,* 14, 191, 1982.
56. **Plant, A. R. and Moore, K. G.,** NADH cytochrome c reductase activity associated with a membrane fraction from protein bodies of *Lupinus angustifolius* cotyledons, *New Phytol.,* 93, 359, 1983.
57. **Viljoen, C. C., Cloete, F., Botes, D. P., and Kruger, H.,** Isolation and characterization of NAD(P)H-dehydrogenases from seeds of the castor bean, *Phytochemistry,* 22, 365, 1983.
58. **Morré, D. J., Vigil, E. L., Frantz, C., Goldenberg, H., and Crane, F. L.,** Cytochemical demonstration of glutaraldehyde-resistant NADH-ferricyanide oxido-reductase activities in rat-liver plasma membranes and Golgi apparatus, *Cytobiologie,* 18, 213, 1978.
59. **Webb, J. L.,** *Enzyme and Metabolic Inhibitors,* Vol. 2, Academic Press, New York, 1966, 670.
60. **Drechsler, Z. and Neumann, J.,** Inhibition of oxygen evolution in chloroplasts by ferricyanide, *Plant Physiol.,* 70, 840, 1982.
61. **Barr, R. and Crane, F. L.,** Ferricyanide reduction in photosystem II of spinach chloroplasts, *Plant Physiol.,* 67, 1190, 1981.
62. **Fitzsimons, J. T. R.,** Enzyme cytochemistry — the present state of the art, *Int. J. Biochem.,* 15, 267, 1983.
63. **Pistorius, E. K., Funkhouser, E. A., and Voss, H.,** Effect of ammonium and ferricyanide on nitrate utilization by *Chlorella vulgaris, Planta,* 141, 279, 1978.
64. **Weavers, B. A.,** An X-ray microanalytical study of the ferricyanide reaction for the electron cytochemical demonstration of succinate dehydrogenase activity in isolated mitochondria, *Histochem. J.,* 6, 121, 1974.
65. **Sawhney, S. K., Naik, M. S., and Nicholas, D. J. D.,** Regulation of NADH supply for nitrate reduction in green plants via photosynthesis and mitochondrial respiration, *Biochem. Biophys. Res. Commun.,* 81, 1209, 1978.
66. **Metzler, D. E.,** *Biochemistry. The Chemical Reactions of Living Cells,* Vol. 2, Academic Press, New York, 1977.
67. **Kalina, M. and Pearse, A. G. E.,** Ultrastructural localization of succinic dehydrogenase and cytochrome oxidase complexes on the mitochondrial crystal membrane; a cytochemical study, in *Microscopie Électronique 1970. Résumés des Comm. Prés. Septieme Congr. Int. Grenoble 1970,* Vol. 3, 133.
68. **Vannoorden, C. J. F. and Tas, J.,** Advantages of 1-methoxy PMS as an electron carrier in dehydrogenase cytochemistry, *Histochem. J.,* 14, 837, 1982.
69. **Kugler, P. and Wrobel, K.-H.,** Meldola blue: a new electron carrier for the histochemical demonstration of dehydrogenases (SDH, LDH, G-6-PDH), *Histochemistry,* 59, 97, 1978.
70. **Kugler, P. A.,** A gel-sandwich technique for the qualitative and quantitative determination of dehydrogenases in the enzyme histochemistry. I. Development of the new methods on the example of LDH (E. C. 1.1.1.27), *Histochemistry,* 60, 265, 1979.
71. **Lippold, H. J.,** Quantitative succinic dehydrogenases histochemistry. A comparison of different tetrazolium salts, *Histochemistry,* 76, 381, 1982.
72. **Butcher, R. G. and Evans, A. W.,** The diffusion of formazans during dehydrogenase reactions, in *Abstr. 6th Int. Histochem. Cytochem. Congr.,* Lake, B. D., Bayliss High, O., Holt, S. J., and Stoward, P. J., Eds., Royal Microscopical Society, Oxford, 1980, 59.
73. **Gahan, P. B. and Kalina, M.,** The use of tetrazolium salts in the histochemical demonstration of succinic dehydrogenase activity in plant tissue, *Histochemie,* 14, 81, 1968.
74. **Spector, G.,** The ultrastructural cytochemistry of lactic dehydrogenase, succinic dehydrogenase, dihydronicotinamide adenine dinucleotide diaphorase and cytochrome oxidase activities in hair cell mitochondria of the guinea pig cochlea, *J. Histochem. Cytochem.,* 23, 216, 1975.
75. **Reiss, J.,** Dimethylsulfoxide as carrier in enzyme cytochemistry, *Histochemie,* 26, 93, 1971.
76. **Yoo, B. Y. and Oreland, L.,** Electron microscopic localization of monoamine oxidase in coleoptiles of *Avena sativum* L. *J. Histochem. Cytochem.,* 23, 784, 1975.
77. **Öpik, H.,** The reaction of mitochondria in the coleoptiles of rice (*Oryza sativa* L.) with diaminobenzidine, *J. Cell Sci.,* 17, 43, 1975.
78. **Gahan, P. B., McLean, J., Kalina, M., and Sharma, W.,** Freezing sectioning of plant tissues: the technique and its use in plant histochemistry, *J. Exp. Bot.,* 18, 151, 1967.

79. **Bullock, G. R.,** The current status of fixation for electron microscopy: a review, *J. Microsc.*, 133, 1, 1984.
80. **Papageorgiou, G. C. and Demosthenopoulou-Karaoulani, E.,** Stabilization of the morphology and the photosynthetic function of isolated intact chloroplasts with glutaraldehyde, *Z. Pflanzenphysiol.*, 105, 201, 1982.
81. **Oku, T., Sugahara, K., and Tomita, G.,** Electron transfer and energy dependent reactions in glutaraldehyde-fixed chloroplasts, *Plant Cell Physiol.*, 14, 385, 1973.
82. **Hajós, F. and Kerpel-Fronius, S.,** The incubation of unfixed tissues for electron microscopic histochemistry, *Histochemie*, 23, 120, 1970.
83. **Auderset, G., Gahan, P. B., Dawson, A. L., and Greppin, H.,** Glucose-6-phosphate dehydrogenase as an early marker of floral induction in shoot apices of *Spinacia oleracea* var. nobel, *Plant Sci. Lett.*, 20, 109, 1980.
84. **Gahan, P. B. and Dawson, A. L.,** Problems encountered with BSPT in dehydrogenase histochemistry, *Histochem. J.*, 13, 338, 1981.
85. **Butcher, R. G., Dawson, A. L., Knaab, S. A., and Gahan, P. B.,** Dehydrogenase activity and loss of formazan from tissue sections, *Histochem. J.*, 12, 591, 1980.
86. **Hayat, M. A.,** Specimen preparation, in *Electron Microscopy of Enzymes. Principles and Methods*, Vol. 1, Hayat, M. A., Ed., Van Nostrand Reinhold, New York, 1973.
87. **Steponkus, P. L.,** Effect of freezing on dehydrogenase activity and reduction of triphenyl tetrazolium chloride, *Cryobiology*, 8, 570, 1971.
88. **Nachlas, M. M., Tsou, K.-C., De Souza, E., Cheng, C.-S., and Seligman, A. M.,** Cytochemical demonstration of succinic dehydrogenase by the use of a new p-nitrophenyl substituted ditetrazole, *J. Histochem. Cytochem.*, 5, 420, 1957.
89. **Fosket, D. E. and Miksche, J. P.,** A histochemical study of the seedling shoot apical meristem of *Pinus lambertiana*, *Am. J. Bot.*, 53, 694, 1966.
90. **Elhiti, M. M. Y., Butler, R. D., and Moore, D.,** Cytochemical localization of glutamate dehydrogenases during carpophore development in *Coprinus cinereus*, *New Phytol.*, 82, 153, 1979.
91. **Wenzel, J. and Behrisch, D.,** Elektronenmikroskopischer Nachweis von Oxydo-Reduktasen im Herzmuskel der Ratte, *Z. Mikrosk.-Anat. Forsch.*, 84, 372, 1971.
92. **Butcher, R. G.,** Studies on succinate oxidation. I. The use of intact tissue sections, *Exp. Cell Res.*, 60, 54, 1970.
93. **Barthová, J. and Leblová, S.,** Influence of light and darkness on the concentration of lactic, glycolic, succinic, malic and citric acid in pea plants, *Biol. Plant.*, 11, 97, 1969.
94. **Borgers, M.,** Enzyme cytochemistry — introductory remarks, in *Electron Microscopy 1980, Vol. 2, Proc. 7th Eur. Congr. on Electron Microscopy*, Brederoo, P. and de Priester, W., Eds., 7th Eur. Congr. Electron Microscopy, Found., Leiden, The Netherlands, 1980, 280.
95. **Naik, M. S. and Nicholas, D. J. D.,** Relation between CO_2 evolution and *in situ* reduction of nitrate in wheat leaves, *Aust. J. Plant Physiol.*, 8, 515, 1981.
96. **Solomonson, L. P.,** Cyanide as a metabolic inhibitor, in *Cyanide in Biology*, Vennesland, E., Conn, E., Knowles, O. J., Westley, J., and Wissing, F., Eds., Academic Press, London, 1981, 11.
97. **Dixon, M. and Webb, E. C.,** *Enzymes*, Longmans, London, 1958.
98. **Eaglesham, A. R. J. and Hewitt, E. J.,** The regulation of nitrate reductase activity from spinach (*Spinacea oleracea* L.) leaves by thiol compounds in the presence of adenosine-5'-diphosphate, *FEBS Lett.*, 16, 315, 1971.
99. **Rathnam, C. K. M. and Das, V. S. R.,** Nitrate metabolism in relation to the aspartate-type C-4 pathway of photosynthesis in *Eleusine coracana*, *Can. J. Bot.*, 52, 2599, 1974.
100. **Redinbaugh, M. and Campbell, W. H.,** Reduction of ferric citrate catalyzed by NADH: nitrate reductase, *Biochem. Biophys. Res. Commun.*, 114, 1182, 1983.
101. **Seckbach, J.,** Ferreting out the secrets of plant ferritin — a review, *J. Plant Nutr.*, 5, 369, 1982.

Specific Enzyme Protocols

ASPARTATE AMINOTRANSFERASE

Rex Paul

INTRODUCTION

Functions

Aspartate aminotransferase (AAT, E.C. 2.6.1.1) is the enzyme that catalyzes the reversible transfer of an amino group between *l*-aspartate and α-ketoglutarate to form oxalacetate and *l*-glutamate. Most AAT cytochemical studies have been performed on animal tissues and have demonstrated presence of the enzyme in mitochondria,[1-4] microbodies,[3] cytoplasm,[5] nucleus,[5] plasma membrane,[6] and in the sarcoplasmic reticulum of muscle.[1,6]

AAT is found ubiquitously in the plant kingdom and has important roles in C_4 photosynthesis, photorespiration, and the breakdown of lipid reserves by glyoxysomes. Not surprisingly, several subcellular locations (cytosol, plastids,[7] mitochondria,[8] peroxisomes,[9] and glyoxysomes[10]) have been described for the enzyme, which correlate with its many metabolic roles. This enzyme, also known as glutamate oxalacetate transaminase and glutamate-aspartate transaminase, is of prime importance in the synthesis and degradation of aspartate and appears to have important functions in seed germination and early embryo development of a variety of plants.[7]

The enzyme has been purified from cauliflower buds, cotton seeds, pea seeds, wheat germ, bushbean roots, and oat and spinach leaves.[7]

Structural Characteristics

Structurally, AAT appears to be of dimeric construction, since heterodimers have been produced in genetic crosses[11] and are found in certain ecotypes.[12] Further, the enzyme contains pyridoxal phosphate as a cofactor.[11] It is known that this coenzyme is the amino group acceptor, forming pyridoxylamine during an intermediate reaction step.[13] Pyridoxal phosphate is thought to be much more tightly bound in plant AAT than in the animal AAT in that the enzyme-coenzyme relationship survives extraction and purification procedures.[7] The purified plant enzyme has been found to have a molecular weight of about 100,000, although the figure varies with the source and procedure.[11] The cytochemical localization of AAT would have application in a wide variety of plant species and tissues. For the purposes of this review, I have chosen tissue from two species which belong to a physiological group in which AAT has an important and well-documented metabolic role.

C_4 Photosynthesis

Plant species are divided into three groups based on their method of assimilating carbon dioxide. These are the C_3 Calvin-cycle types (C_3 plants), the C_4-dicarboxylic acid cycle types (C_4 plants), and the crassulacean acid metabolism types (CAM plants).

C_4 plants are divided into three groups (subtypes) on the basis of the enzyme catalyzing the release of carbon dioxide in the bundle sheath cells.[14] The three divisions include the $NADP^+$ malic enzyme (NADP-ME), the NAD-malic enzyme (NAD-ME), and the phosphoenol pyruvate carboxykinase (PCK) groups. In each of the three groups, CO_2 is incorporated into C_4 dicarboxylic acids in mesophyll cells and transported into usually adjacent bundle sheath cells. Here the C_4 acids are decarboxylated and the CO_2 is incorporated into the Calvin cycle. In the $NADP^+$-ME group, malate is the predominant C_4 acid transported to the bundle sheath cell for decarboxylation. In the latter groups, aspartate is transported. The performance of this shuttle requires that AAT must be active in both bundle sheath and mesophyll cells in order for photosynthesis to proceed.

RATIONALE FOR ASPARTATE AMINOTRANSFERASE CYTOCHEMICAL LOCALIZATION

The basis for the cytochemical localization of aspartate aminotransferase involves providing the enzyme with its two requisite substrates, l-aspartate and α-ketoglutarate in a medium of lead (Pb^{2+}) ions.[15] The lead ions have been shown not to inhibit the enzyme at the concentration used in the incubation medium.[16] The reaction is an example of a simultaneous capture mechanism[17] and can be diagrammed as follows:

$$l\text{-aspartate} + \alpha\text{-ketoglutarate} \xrightleftharpoons[\text{aminotransferase}]{\text{aspartate}}$$

$$\text{glutamate} + \text{oxalacetate}$$
$$\Downarrow Pb^{++}$$
$$\text{lead oxalacetate}$$
$$\text{(Electron dense)}$$

As is the case in many enzyme cytochemical localizations, the procedure for AAT involves an initial short aldehyde fixation prior to incubation in the reaction or control medium. This step must stabilize (insolubilize) the enzyme without denaturing it.[18] The tissue is then reacted in the incubation medium (or control medium), washed in buffer, postfixed in OsO_4, dehydrated (e.g., in cold acetone series), and then embedded (e.g., in Spurr's[19] medium). Each of these steps is detailed in the appropriate following section.

FIXATION

Aldehyde Fixation

Initial fixation is a very critical step to the quality of the localization, in that the intracellular position of AAT must be stabilized without destroying its activity. Exposure to aldehyde fixatives has the further effect of facilitating the diffusion of incubation medium into the tissue.

The effect of aldehyde fixatives on the activities of animal[16,18,20] and plant[21] aspartate aminotransferases has been investigated. It was found that aldehyde fixatives rapidly inactivate the enzyme. By carefully timing exposure to aldehyde fixatives, particularly glutaraldehyde, the optimal compromise between tissue preservation and enzyme inactivation can be determined.

Substrate Protection

Papadimitriou and Van Duijn[20] have shown that the addition of α-ketoglutarate to the fixation medium allows more exposure of the enzyme to the fixative before it is inactivated. Presumably, the α-ketoglutarate covers active sites in the enzyme, prolonging the time required for inactivation by the fixative. This procedure, called substrate protection, prolonged AAT activity, particularly in the cytosolic isozyme, which is the most vulnerable to inactivation by glutaraldehyde. Exposure to glutaraldehyde, at least in plant tissue, is necessary for adequate tissue fixation.

Fixation Schedule

Trials by Fomina et al.[21] and by Vaughn and Paul[12] have shown that a short (5 min) exposure to cold (4°C) glutaraldehyde, followed by treatment with cold paraformaldehyde, adequately fixed plant tissue without destroying AAT enzyme activity. AAT activity that survives glutaraldehyde fixation in animal tissue is not greatly affected by up to 30 min of subsequent formaldehyde fixation that would, in the absence of previous glutaraldehyde exposure, denature it.[18] Lee and Torack[2] proposed that fixation survivability of the enzyme depends on the physicochemical orientation of the protein within membranes. The activity of the purified solubilized enzyme is destroyed by aldehyde fixatives.

Schedule	Time
2% Glutaraldehyde in 0.2 M cacodylate buffer pH 7.6 (117 mg of α-ketoglutarate may be added per 100 mℓ for substrate protection)	5 min
4% Paraformaldehyde in 0.2 M cacodylate buffer pH 7.6 (117 mg of α-ketoglutarate may be added per 100 mℓ for substrate protection)	30 min
Rinse in 0.2 M cacodylate buffer pH 7.6 (117 mg/100 mℓ of α-ketoglutarate may be added per 100 mℓ in first two changes of buffer); total of six changes of buffer over 1—2 hr; final change is made at room temperature	1—2 hr

The 5-min fixation in 2% glutaraldehyde[21] corresponds to the perfusion fluid fixation of Lee,[13] which is 1% glutaraldehyde and 4% formaldehyde in 0.25 M sucrose. Sucrose in the fixation and subsequent media may badly plasmolyze plant tissue and is not required for adequate substrate penetration and prevention of mitochondrial swelling, as apparently is the case in animal tissue. The plant tissue, after being transferred from glutaraldehyde to paraformaldehyde, may be further minced (<0.5 mm^3) for subsequent paraformaldehyde fixation.

INCUBATION MEDIUM

The AAT cytochemical medium of Lee[13] for animal tissue consists of

l-Aspartate	20 mM
α-Ketoglutarate	2—4 mM (Depending on reaction speed desired,[13] 4 mM optimal for localization of the soluble isozyme[20])
Lead nitrate	6 mM
Imidazole	50 mM
Sucrose	0.25 M

This formula approximates that of Fomina et al.[21] for plants. In the latter incubation media, the lead nitrate concentration is 3 mM, and the sucrose is omitted.

Formula — Add to 50 mℓ H$_2$O (boiled, distilled) 133.1 mg l-aspartate, 29.2 mg α-ketoglutarate, and 17.0 mg imidazole. Adjust pH to 7.6 with 0.4 N NaOH.

The pH optimum for most plant aspartate aminotransferases appears to be 8.0 to 8.5.[1] Hatch[22] reported the mitochondrial enzyme of *Atriplex* to have a pH optimum of 8.0 to 8.5, whereas the mesophyll isozyme is very active near neutrality.

Next, add slowly with stirring 49.7 mg Pb(NO$_3$)$_2$.

The presence of both aspartate and imidazole is required for the chelation of the lead ions and the prevention of their precipitation by α-ketoglutarate at the slightly alkaline pH required for this reaction.[13,15] One product, oxalacetate formed in the reaction, does precipitate lead under these conditions. This precipitate, lead oxalacetate, is redissolvable at acid pH.[15]

The procedure of Lee[13] utilizes two solutions, one with 40 mM l-aspartate, 4 to 8 mM α-ketoglutarate, and 100 mM imidazole, and the second with 12 mM lead nitrate. They are mixed in equal volumes just prior to incubation. This increases shelf life, as the separate solutions can be stored in the cold for at least 1 week.[13]

Controls

Lee[13] suggests two controls, one with d-aspartate substituted for l-aspartate in the reaction medium, and the other with α-ketoglutarate omitted. The latter control must be used with great care if substrate protection is employed, especially if the protectant, α-ketoglutarate, is included in the postfixation rinse. At least two final rinses should be made excluding α-ketoglutarate from the solution. Fomina et al.[21] utilized incubation medium with α-ketoglutarate omitted as one control as well as the addition of an inhibitor (translated from Russian as "Tubazid", probably Isoniazid) to the incubation medium as a second control. Two other aspartate aminotransferase inhibitors, 2-aminooxyacetic acid and 2-amino-4-methoxy-3-butenoic acid, have been shown to irreversibly react with the coenzyme.[23] The latter inhibitor binds to the apoprotein as well.[23] These could be included in the reaction medium as a control provided they do not react with other components.

The incubation procedure may be performed at room temperature with constant agitation[13] (or at 40°C with no agitation[18]).

Wash

The postincubation wash is carried out in buffer at pH 7.6 to preserve the insolubility of lead oxalacetate. The pH should never be allowed to become more acidic once the reaction product is formed, since acidic pH dissolves lead oxalacetate.[15] Lee[13] leaves l-aspartate (or d-aspartate in the d-aspartate control) in the wash solution to facilitate chelation and removal of loosely bound lead left in the tissue.

POSTFIXATION

The tissue, after being washed, is postfixed overnight in 1% OsO_4 in 0.1 M cacodylate buffer, pH 7.6, at 4°C. Demel and Jozsa[1] add 1% uranyl acetate to 1% OsO_4 in collidine buffer, pH 7.3. Such *in bloc* staining, however, must be used with care, since it may obscure the localization in plant tissue.

The next step, the wash, is not done in distilled water, but in cacodylate buffer, pH 7.6, at 4°C for 1 to 2 hr.

DEHYDRATION AND EMBEDDING

Lee and Torack[3] found that dehydrating the tissue in buffered ethanol resulted in more precise localization of the product. This was due to the maintenance of alkalinity and, thus, the insolubility of intracellular deposits of lead oxalacetate. We[12] utilized this principle, but found acetone to be the superior dehydration fluid in plant tissue.

Dehydration is carried out in a graded series of cold (4°C) cacodylate buffer, pH 7.6, and acetone. The dehydration fluid becomes cloudy in the higher acetone concentrations (e.g., 90% acetone, 10% buffer), but this apparently does not affect the tissue. When a concentration of 100% cold acetone is reached, the vials are allowed to warm to room temperature. One final change of 100% acetone at room temperature is made before embedding in Spurr's[19] medium.

Once polymerized blocks are obtained, thin sections should be taken from the very edge of the embedded tissue. Interior regions of the tissue block are usually unstained and poorly fixed.

RESULTS

The tissues chosen for this cytochemical study were taken from two NAD-ME-type C_4 species, one monocotylededonous species *Cynadon dactylon* (L.) Pers., and one dicotyle-

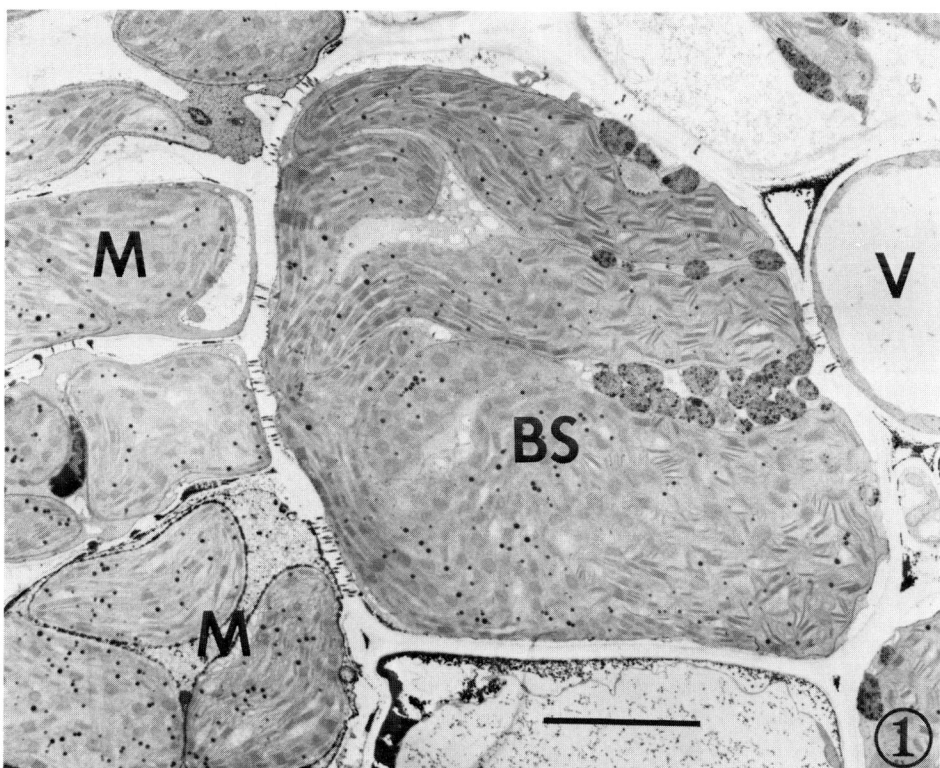

FIGURE 1. Portion of vascular bundle of *Cynodon dactylon* showing reaction product in mesophyll (M) cell cytoplasm, mitochondria, and periphery of chloroplasts. Reaction product in bundle sheath cells (BS) is primarily located in mitochondria. V = vascular tissue. Bar = 5 μm.

donous species, *Amaranthus hybridus* L. This C_4 subtype was chosen because of the contribution of aspartate aminotransferase to photosynthesis in these species. Furthermore, the AAT-containing bundle sheath mitochondria in these species possess very extensive cristae membrane systems.[14]

An overall view of AAT localization within the bundle sheath of *Cynodon dactylon* is shown in Figure 1. The mitochondria, chloroplast envelopes, and general cytoplasmic areas within mesophyll cells stain positively for AAT. The plasmodesmata between mesophyll and bundle sheath and between bundle sheath and vascular tissue are normally osmiophillic, but do appear to stain for AAT. Some external cell wall surfaces appear to stain positively for AAT. The most consistently staining structures are the mitochondria of bundle sheath cells. This probably reflects the situation in rat liver where fixed mitochondria retained 80% of their AAT activity.[2] This was thought to be due to membrane fixation occurring before any protein denaturation effects, thus, stabilizing the enzyme *in situ*. Leaf tissue of *Cynodon dactylon* incubated in *d*-aspartate control media (Figure 2) shows little lead oxalacetate deposition.

A more highly magnified view of a *Cynodon dactylon* mesophyll (Figure 3) reveals the pattern of lead oxalacetate deposition. The cytoplasm in this example is densely stained at certain locations within the cell. Mitochondria (Figure 3) stain in the cristae spaces as well as at their borders, while mesophyll chloroplasts stain for AAT near their envelopes. The AAT isozymes associated with cytoplasm are known to be very soluble so that interpretation as to location of cytoplasmic AAT reactions must be carefully made.

In the bundle sheath, mitochondria stained for AAT (Figure 4) appear to have activity

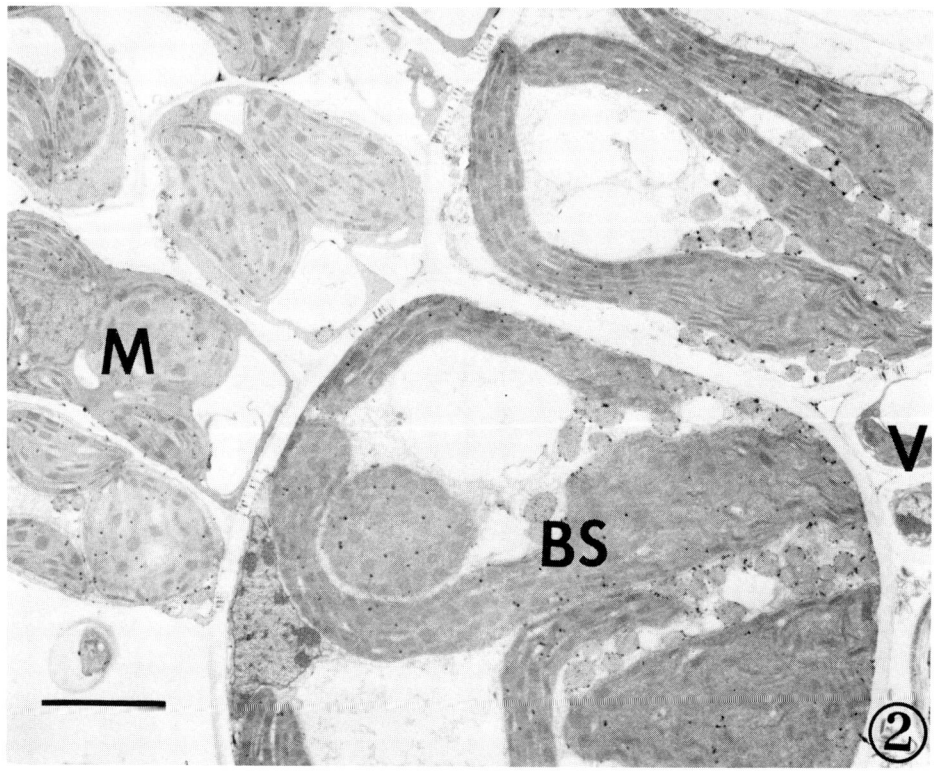

FIGURE 2. Portion of *Cynodon dactylon* leaf as in Figure 1 except that if is *d*-aspartate control. There appears to be no nonspecific lead deposition. Bar = 5 μm.

primarily in the mitochondrial matrix. In heavily stained regions in these organelles, cristae are clearly seen (Figure 4, arrows). Microbodies (Figure 4) stain only at the periphery. Bundle sheath cells incubated in *d*-aspartate control medium (Figure 5) have little mitochondrial or microbody AAT stain. The mitochondria do, however, have small osmiophillic structures associated with their surfaces.

Amaranthus hybridus leaf tissue (Figures 6 and 7) has much the same pattern of lead oxalacetate deposition as *Cynodon*. This underscores the similarities of physiology, especially in function and intracellular location of aspartate aminotransferase among these otherwise very different, unrelated species.

The success of this cytochemical reaction depends upon the reactivity of oxalacetate with chelated lead ions to form an insoluble electron-dense product. The maintenance of lead in the reaction medium is dependent upon the presence of aspartate and imidazole as interacting lead-chelating ingredients.[13] The extention of this cytochemical reaction to other transaminases would rely on the successful substitution of the corresponding substrate for aspartate in the lead-chelating incubation medium. Lee[13] states that if aspartate is replaced by either glutamate or alanine in the incubation medium, the lead will precipitate when the pH is raised near neutrality. This factor may prohibit the extension of this cytochemical technique to other aminotransferases.

The cytochemical localization of AAT should find application in metabolic studies of a wide variety of plant tissues. Future applications may include investigations into seed germination, lipid metabolism, and C_3, CAM, and C_4 photosynthesis.

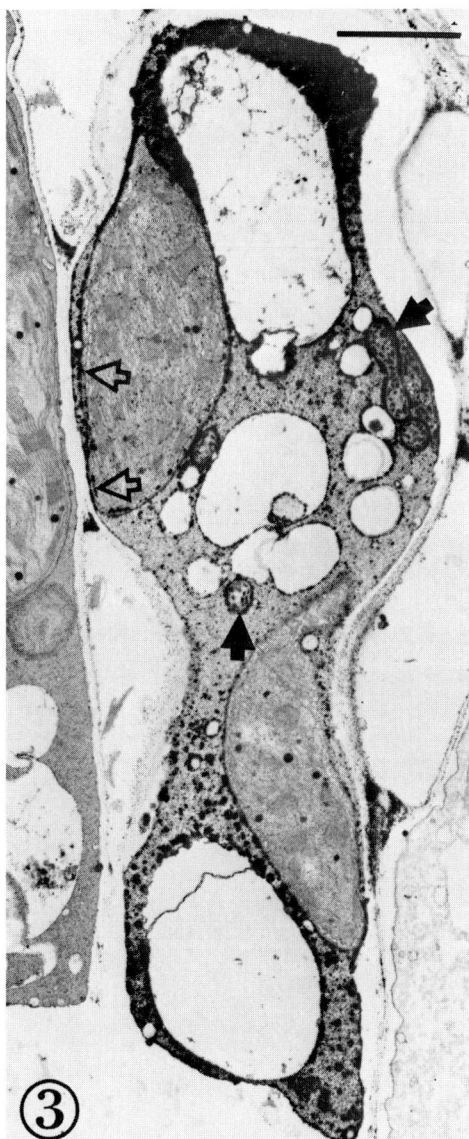

FIGURE 3. Mesophyll cell of *Cynodon dactylon*. Cytoplasm is densely stained at both ends of elongated cell. Mitochondria (arrows) show deposits within cristae. The plastid peripheral reticula (hollow arrows) appear to contain some reaction product. Bar = 2 μm.

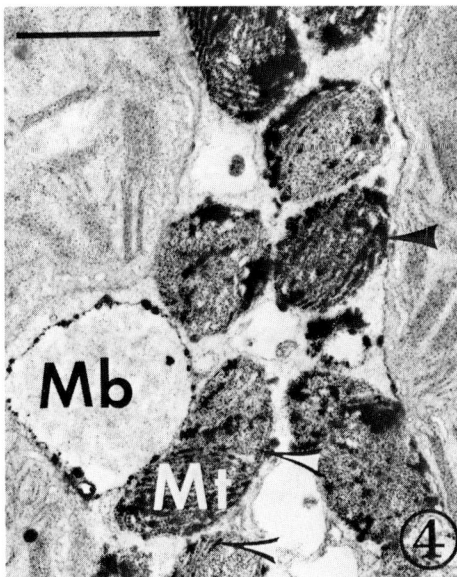

FIGURE 4. Bundle sheath cell of *Cynodon dactylon* showing lead oxalacetate deposits in mitochondria (Mt) and around periphery of microbody (Mb). Bar = 1 μm.

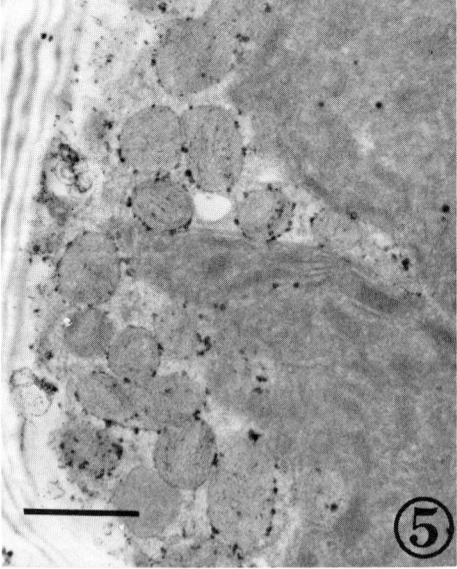

FIGURE 5. Bundle sheath cell of *Cynodon dactylon* with *d*-aspartate substituted for *l*-aspartate in the reaction media. Bar = 2 μm.

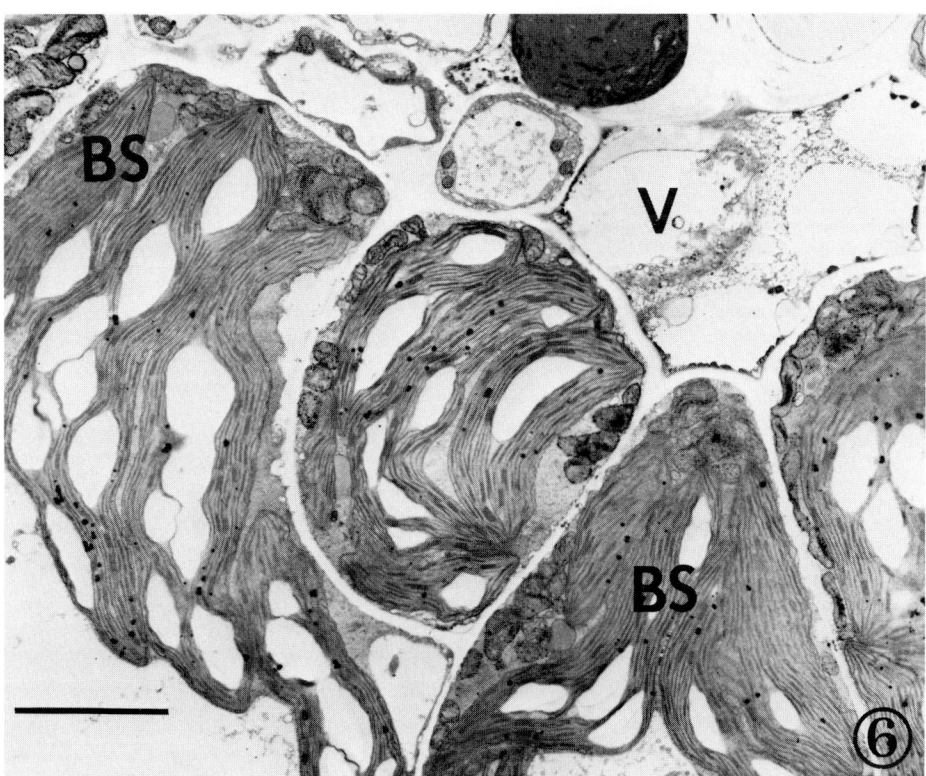

FIGURE 6. Bundle sheath cells of *Amaranthus hydridus* showing positive staining for aspartate aminotransferase in mitochondria. V = vascular tissue. Bar = 5 μm.

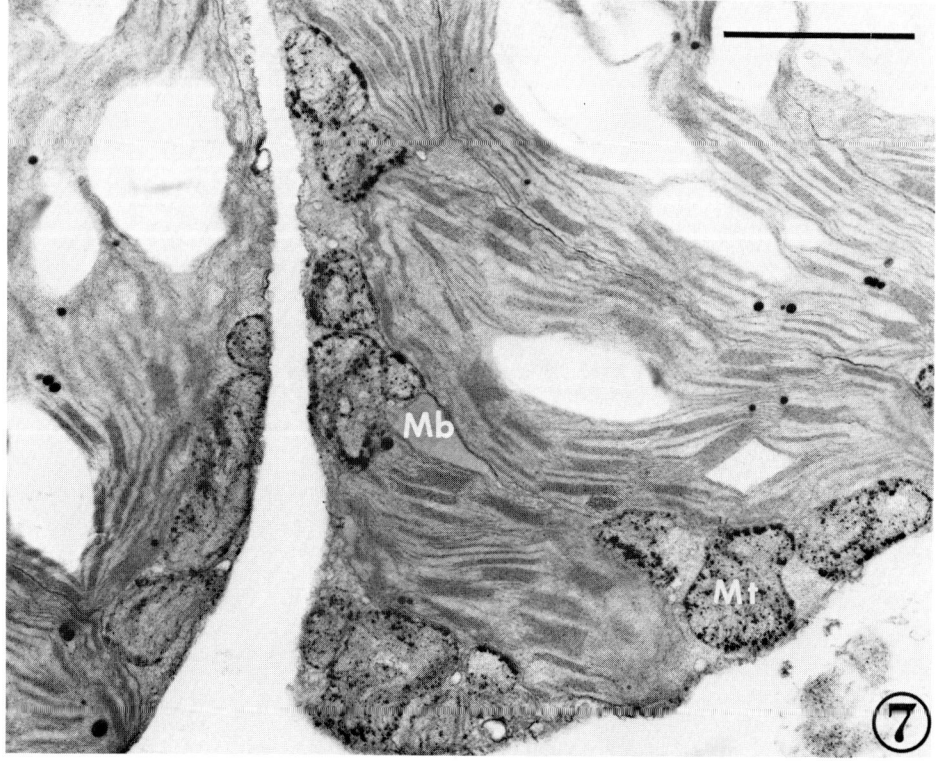

FIGURE 7. Enlarged view of bundle sheath cells of *Amaranthus hydridus* stained for aspartate aminotransferase. Note staining in mitochondria (Mt) and lack of stain in the pliable microbody (Mb). Bar = 2 μm.

REFERENCES

1. **Demel, S. and Jozsa, L.**, Ultrahistochemical localization of the glutamate-oxalacetate transaminase (GOT) enzyme in hepatic and skeletal muscle cells in rat, *Acta Histochem.*, 72, 27, 1983.
2. **Lee, S. H. and Torack, R. M.**, A biochemical study of glutamic oxalacetic transaminase activity of rat hepatic mitochondria fixed in situ and in vitro, *J. Cell Biol.*, 39, 725, 1968.
3. **Lee, S. H. and Torack, R. M.**, Electron microscope studies of glutamic oxalacetic transaminase in rat liver cell, *J. Cell Biol.*, 39, 716, 1968.
4. **Lee, S. H.**, Ultrastructural localization of glutamic oxalacetic transaminase activity in cardiac muscle fiber and cardiac mitochondrial fraction of the rat, *Histochemie*, 19, 99, 1969.
5. **Lee, S. H.**, The possible role of the vesicles in renal ammonia excretion: an implication of concentrated glutamic oxalacetic transaminase, *J. Cell Biol.*, 45, 644, 1970.
6. **Papadimitrious, J. M. and Van Duijn, P.**, The ultrastructural localization of the isozymes of aspartate aminotransferase in murine tissues, *J. Cell Biol.*, 47, 84, 1970.
7. **Wightman, F. and Forrest, J. C.**, Properties of plant aminotransferases, *Phytochemistry*, 17, 1455, 1978.
8. **Hatch, M. D. and Mau, S.-L.**, Activity, location, and role of aspartate aminotransferase and alanine aminotransferase isoenzymes in leaves with C_4 pathway photosynthesis, *Arch. Biochem. Biophys.*, 156, 195, 1973.
9. **Rehfeld, D. W. and Tolbert, N. E.**, Aminotransferases in peroxisomes from spinach leaves, *J. Biol. Chem.*, 247, 4803, 1972.
10. **Cooper, T. G. and Beevers, H.**, Mitochondria and glyoxysomes from castor bean endosperm, *J. Biol. Chem.*, 244, 3507, 1969.
11. **Givan, C. V.**, Aminotransferases in higher plants, in *The Biochemistry of Plants*, Vol. 5, Stumpf, P. K. and Conn, E. E., Eds., Academic Press, New York, 1980, 329.

12. **Vaughn, K. C. and Paul, R. N.**, Localization of aspartate: 2-oxoglutarate aminotransferase in a NAD-ME C_4 species, in preparation.
13. **Lee, S. H.**, Glutamate oxalacetate transaminase, in *Electron Microscopy of Enzymes*, Vol. 1, Hayat, M. A., Ed., Van Nostrand Rheinhold, New York, 1973, 116.
14. **Hatch, M. D., Kagawa, T., and Craig, S.**, Subdivision of C_4 pathway species based on differing C_4 acid decarboxylating systems and ultrastructural features, *Aust. J. Plant Physiol.*, 2, 111, 1975.
15. **Lee, S. H.**, Histochemical demonstration of glutamic oxalacetic transaminase, *Am. J. Clin. Pathol.*, 49, 568, 1968.
16. **Lee, S. H. and Torack, R. M.**, The effects of lead and fixatives on activity of glutamic oxalacetic transaminase, *J. Histochem. Cytochem.*, 16, 181, 1968.
17. **Sexton, R. and Hall, J. L.**, Enzyme cytochemistry, in *Electron Microscopy and Cytochemistry of Plant Cells*, Hall, J. L., Ed., Elsevier/North-Holland, Amsterdam, 1978, 63.
18. **Lee, S. H. and Torack, R. M.**, Aldehyde as fixative for histochemical study of glutamic oxaloacetic transaminase, *Histochemie*, 12, 341, 1968.
19. **Spurr, A. R.**, A low viscosity epoxy resin embedding medium for electron microscopy, *J. Ultrastruct. Res.*, 26, 31, 1969.
20. **Papadimitriou, J. M. and Van Duijn, P.**, Effect of fixation and substrate protection on the isoenzymes of aspartate aminotransferase studied in a quantitative cytochemical model system, *J. Cell Biol.*, 47, 71, 1970.
21. **Fomina, I. R., Bil, K. Y., and Magomedov, I. M.**, Histochemical determination of the localization of aspartate aminotransferase in autotrophic tissues of leaves of the C_4 plant *Amaranthus lividus*, *Dokl. Akad. Nauk SSSR*, 256, 1514, 1981.
22. **Hatch, M. D.**, Separation and properties of leaf aspartate aminotransferase and alanine aminotransferase isoenzymes operative in the C_4 pathway of photosynthesis, *Arch. Biochem. Biophys.*, 156, 207, 1973.
23. **Balkow, C. and Wildner, G. F.**, Aspartate aminotransferases of *Panicum miliaceum* L. and *Panicum antidotale* Retz. Inactivation and reconstitution, *Planta*, 154, 477, 1982.

ENZYME CYTOCHEMISTRY AND IMMUNOCYTOCHEMISTRY OF NUCLEASES

David H. Clapham

INTRODUCTION AND TERMINOLOGY

Nuclease terminology is complex.[1-3] Briefly, "nuclease" is often applied to any enzyme that hydrolyzes a phosphodiester bond in a nucleic acid (DNA or RNA). More strictly, a nuclease acts on *both* DNA and RNA; a DNase acts only on DNA, an RNase only on RNA. *Endonucleases* break bonds in the middle of duplex or single-stranded DNA or RNA. *Exonucleases* work from the ends of molecules; an individual enzyme detaches mono- or oligonucleotides either from the 5' or 3' end, rarely from both. Endo-exonucleases are also known.

The products of nuclease action are nucleotides ending either in 5'-phosphate or 3'-phosphate (Figure 1). No known enzyme produces both 3'- and 5'-nucleotides. In many cases, mononucleotides are released sooner or later, but some endonucleases break the nucleic acid in relatively few places so that the products are oligo- or polynucleotides (e.g., mammalian DNase I) or long double-stranded DNA molecules (e.g., bacterial restriction enzymes).

Although all nucleases could reasonably be called *phosphodiesterases,* this term is reserved for enzymes that degrade small artificial molecules such as α-naphthyl thymidine 3' (or 5')-phosphate, on which most nucleases are inactive. Cyclic nucleotide phosphodiesterases are not considered here, nor are glycosylases that excise foreign bases from DNA.

Nucleases are required at many points in nucleic acid metabolism: not just "restriction" and digestion inside and outside cells, but in at least four distinct steps of DNA replication, in the repair and recombination of DNA, and in the maturation of messenger, ribosomal, and transfer RNA. There is a correspondingly long list of nucleases known in bacterial and mammalian cells, varying from unspecific enzymes that break down polynucleotides to mononucleotides without regard to base sequence, secondary structure, or type of sugar, to highly specialized enzymes such as those that make a single break to one side of a specific lesion in duplex DNA.[4] Enzyme connoisseurs appreciate the elegance and subtlety of nuclease action.[5]

Little is known of plant nucleases. Wilson[2] named and described the following.

- *RNase I* is a soluble endoribonuclease that releases 2':3'-cyclic nucleotides, with a preference for guanosine cyclic monophosphate as a major early product. Only purine cyclic nucleotides are hydrolyzed. The pH optimum is near 5. The molecular weight is 20,000 to 25,000.
- *RNase II* is a microsomal endoribonuclease that releases 2':3'-cyclic nucleotides, with a preference for guanosine cyclic monophosphate as the major early product. Both purine and pyrimidine cyclic monophosphates are hydrolyzed, with a preference for purine cyclic monophosphates. The pH optimum is near 6. The molecular weight is 17,000 to 20,000.
- *Nuclease I* is a particle-bound sugar-nonspecific endonuclease that releases 5'-mononucleotides with a preference for (deoxy)adenosine 5'-mononucleotides as the major early product. It is about equally active on RNA and denatured DNA, but is much less active on native DNA. The pH optimum is near 6. The molecular weight is about 33,000. The enzyme is usually inhibited by EDTA and often requires a divalent cation for maximum activity or stability. The nuclease is accompanied by a 3'-nucleotidase activity.

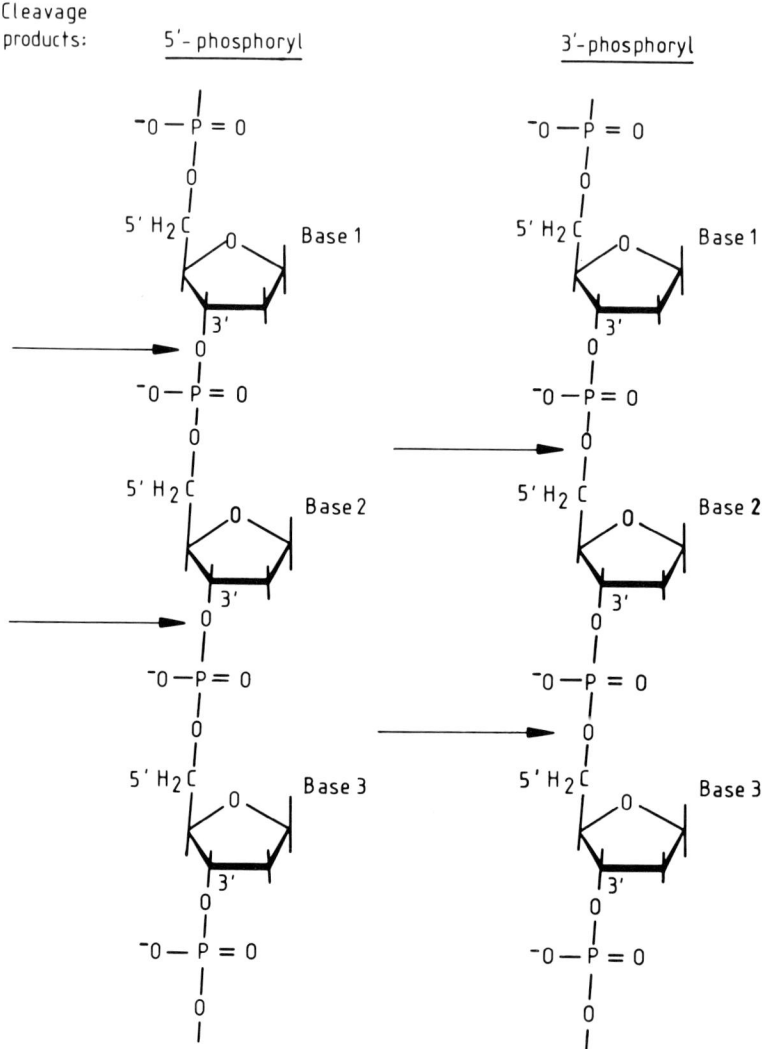

FIGURE 1. Nucleases cleave polynucleotide chains to yield either 5'-phosphoryl or 3'-phosphoryl ends. (After Kornberg, A., *DNA Replication*, Freeman, San Francisco, 1980, chap. 10. With permission.)

- *Plant exonucleases*[2,3] hydrolyze polynucleotide chains from the 3' end releasing 5'-mononucleotides. The pH optimum is 7 to 9. The molecular weight is 100,000 or above. They have been assayed by their action on artificial phosphodiesterase substrates and broadly resemble venom phosphodiesterases.

Wilson considered that there was no sufficiently well-characterized plant enzyme to be named DNase I.

There are, undoubtedly, many more plant nucleases awaiting discovery or fuller description. At present botanists should follow Wilson's nomenclature; it should soon be possible to designate a plant DNase I and perhaps II and III,[6-9] for example. Trivial synonyms for Wilson's term "nuclease I", such as "major plant endonuclease" or the ambiguous "nuclease", should be avoided, even if more recent papers[10,11] suggest that the original working description requires modification. The Enzyme Commission's fine but awkward system of nomenclature was not followed in a recent monograph.[12]

None of the enzymes RNase I, RNase II, and nuclease I has been localized by cell fractionation. Valuable recent papers in which vacuoles have been prepared from plant protoplasts show that a major part of the phosphodiesterase activity (probably due to Wilson's exonuclease) is found in the vacuoles of castor bean[13] and of cultured tobacco cells and tulip petals.[14] A major part of the RNase activity was also found in the vacuoles, but this could be attributed to any or all of at least four enzymes. It is clear that there is plenty of opportunity for cytochemical studies of nucleases, especially in correlative studies with cell fractionation techniques.

The cytochemistry of nucleases has attracted few biologists of any kind and extremely few botanists. This chapter reviews the various proposals for enzymatic methods, illustrated with botanical examples where possible. Then the currently more promising field of nuclease immunocytochemistry is considered. Nuclease cytochemistry deserves more workers, in view of the diversity of the enzymes, their distribution over most parts of most cells, and their importance in so many areas of nucleic acid metabolism.

ENZYME CYTOCHEMISTRY OF NUCLEASES

Substrate Film Method
Principle

A gelatine film containing substrate (DNA or RNA) is placed in contact with tissue sections. After a period of incubation, the gelatine film is removed from the slide containing the sections and treated with a nucleic acid stain such as toluidine blue. Unstained parts of the gelatine gel correspond to regions of nuclease activity in the sections.[15,16]

Results and Discussion

The method has chiefly been applied to animal material.[15-20] In a botanical study, RNase was localized in pollen grain walls,[21] confirming results using a lead precipitation method. The substrate film method is the most direct means available for the histochemical demonstration of nuclease activity; but the low resolution restricts studies to tissue rather than subcellular distribution of enzyme. For further discussion see Daoust's reviews.[16,18]

Lead Precipitation Method with Exogenous Phosphatase
Principle

Semithin or thick sections are incubated in a medium at pH 5.0 to 6.0 containing (1) DNA or RNA, (2) acid phosphatase, and (3) Pb^{2+} ions from lead nitrate or other suitable lead salt.[22,23]

In successful cases, sites in the section containing nuclease hydrolyze the nucleic acid to short nucleotides. Then phosphate ions are immediately split off from the nucleotides by the exogenous phosphatase and precipitated very close to the site of the nuclease, as the highly insoluble lead phosphate. For light microscopy, the sections are further treated with hydrogen sulfide or ammonium sulfide to convert the white lead phosphate to the more visible black lead sulfide.[24] For electron microscopy, the thick sections (30 to 100 μm) are osmicated and embedded and ultrathin sections cut as usual.

For the localization of alkaline nucleases, the incubation medium can be buffered at pH 8.0 and the acid phosphatase replaced by alkaline phosphatase.[25]

An Example

Recent work[26] describing the localization of DNase in pollen can be considered in more detail. Anthers of *Chlorophytum comosum* 1 to 2 hr before anthesis were chopped into fine pieces in a drop of cold 3% glutaraldehyde in 0.1 *M* cacodylate buffer, pH 7.2, and then fixed for 1 hr at 0 to 4°C in a larger volume of the same solution. Material was washed in

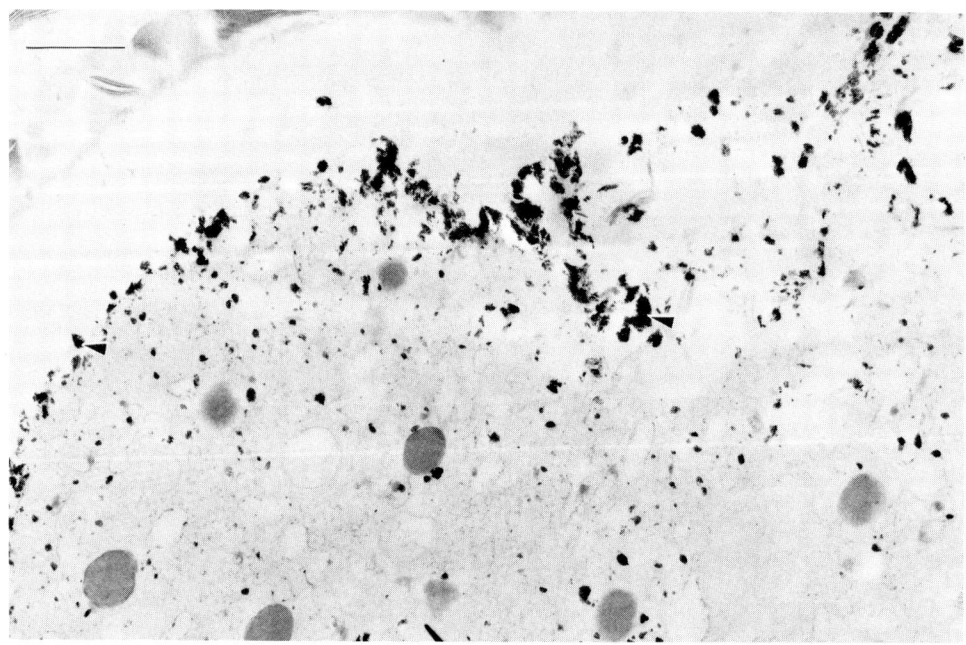

FIGURE 2. Vesicles with precipitated lead phosphate (arrow) marking sites of DNase activity around the vegetative cell wall of the mature pollen of *Chlorophytum comosum*. Bar = 1.0 μm. (From Vaughn, K. C., *Histochemistry*, 73, 487, 1982. With permission.)

four changes of 0.05 M acetate buffer, pH 5.0, at 0 to 4°C over 6 hr and transferred to the following reaction medium:[26,27]

DNA, low molecular weight from herring sperm (Sigma type IV)	0.4% (w/v)
Acid phosphatase, from potato (Sigma type IV)	2 mg/mℓ
Lead nitrate dissolved in 0.05 M acetate buffer, pH 5.0, made from freshly boiled twice-distilled water to prevent precipitation of lead carbonate	10 mM

The reaction medium was allowed to penetrate for 18 hr at 0 to 4°C, then the material was placed in fresh medium at 37°C. After 90 min the material was washed in cacodylate buffer, refixed for 2 hr in 1% OsO_4 in cacodylate buffer at 0 to 4°C, and embedded and sectioned.

As controls, anther sections were boiled or were incubated in the absence of DNA, phosphatase, or lead nitrate, when at most some small, nonspecific products were detected.

Vaughn concluded that DNase activity was localized in vesicles along the intine (near the aperture where the pollen tube would emerge, in particular) and around the generative cell: see Figure 2. The study is of interest in connection with the uniparental transmission of mitochondria, which, perhaps, requires the breakdown or alteration of DNA in the mitochondria of the generative cell.

Discussion

First, is the nuclease activity accurately located in the cell? Second, what nuclease activity is being measured, since many enzymes hydrolyze phosphodiester bonds in DNA or RNA?

Gomori's lead precipitation method for phosphatases is discussed elsewhere in this handbook. Despite various theoretical objections and practical pitfalls, phosphatase can be ac-

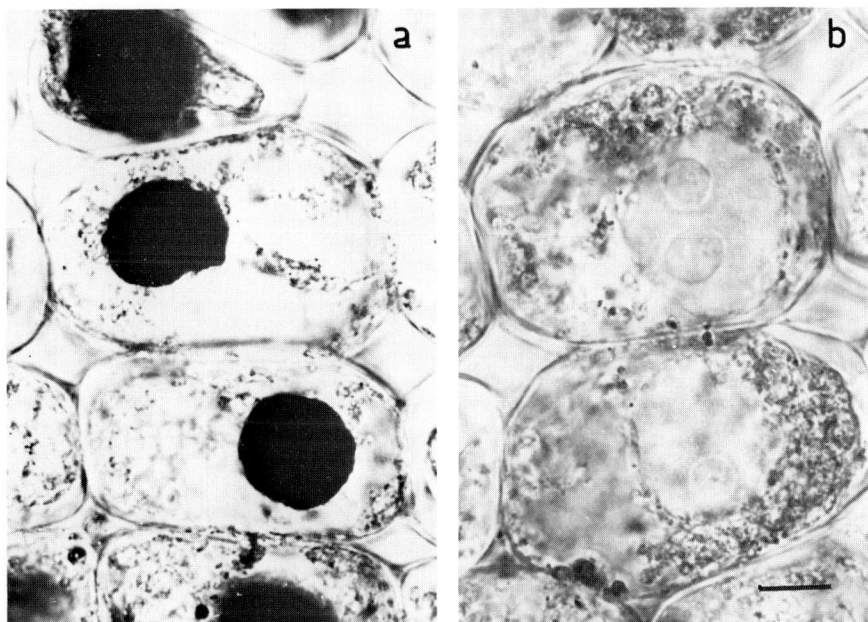

FIGURE 3. Cortical cells from Vibratome sections of *Tradescantia* root tips incubated for 16 hr at 37°C in DNase reaction media containing acid phosphatase from wheat (a) or from potato (b). Nuclei are stained in (a), not in (b). Bar = 10 μm. (From Clapham, D., *Protoplasma*, 107, 109. 1981. With permission.)

curately localized in thin and ultrathin sections of animal and probably also plant material. The extension of Gomori's method to nuclease introduces an extra contentious step, the addition of acid phosphatase to the reaction medium. The external acid phosphatase may well be capable of splitting off phosphate from newly formed mono- and oligonucleotides sufficiently rapidly to allow accurate localization of tissue nuclease, but this needs experimental support for each tissue studied.

A further point concerns the purity of the acid phosphatase preparation. Most workers — though not Vaughn[27] — have used a commercial preparation from wheat germ (e.g., Sigma type I). This extremely crude preparation contains much nuclease (!) active on DNA and RNA at pH 5.0 and 6.0.[28] The nuclease breaks down DNA in the reaction medium to nucleotides that can be acted on by tissue phosphatases and nucleotidases; or, alternatively, by the phosphatase in the reaction medium, leading to the formation of lead phosphate and its absorption by the tissue. That this is not unimportant may be seen by comparing Figure 3a and b. Potato acid phosphatase (Sigma type IV) is nearly free of nuclease.[28]

A positive reaction in the absence of added phosphatase may indicate a site of phosphatase in the cell; this may or may not be the same site as that of the nuclease, depending on how far the released nucleotides diffuse.

It is important to run a control with heat-inactivated sections, not least because this can detect consequences of impure phosphatase.[28] This control has often been omitted, as in studies of DNase in *Cucurbita*[29] and *Vicia*[30] roots and of RNase in pollen.[21] The authors' conclusions may, nevertheless, have been correct, particularly in the pollen study, where localization was checked by the substrate film method.

What nuclease activity is measured (at best) by the lead precipitation method? This depends on the substrate employed. If nucleases active on DNA are to be studied, the substrate DNA may be double or single stranded and of high or low molecular weight, or a mixture of these. In order to facilitate entry of substrate into the chopped anthers, Vaughn[26,27] used a preparation of degraded herring sperm DNA, of mostly low molecular weight, and probably both double and single stranded. Many nucleases, including phosphodiesterases, probably act on such DNA. If semithin sections (10 μm or less) are incubated in the reaction medium, then high molecular weight native or single-stranded DNA from calf thymus or salmon sperm can be used as substrate rather than degraded DNA.[25] Phosphodiesterases have little activity on native high molecular weight DNA, in contrast to enzymes such as pancreatic DNase (mammalian DNase I) and acid DNase (mammalian DNase II). Again, enzymes such as plant nuclease I are very active on high molecular weight single-stranded DNA, such as heat-denatured salmon DNA, but little active on native salmon DNA. Possibilities of distinguishing different types of DNase cytochemically have not been explored for plant material. In cases where the products of DNase action are oligo- rather than mononucleotides, it is clearly important that the exogenous phosphatase is active on oligonucleotides, as are prostatic acid and bacterial alkaline phosphatase,[31] not just on mononucleotides.

Conclusions

Despite shortcomings, the lead precipitation method with exogenous phosphatase is sufficiently promising to merit further critical botanical study, for example, along the lines of Taper,[25] who used animal material. It is too early to damn the method owing to the undoubtedly weak assumptions underlying the addition of phosphatase to the reaction medium. In future work, points to consider include the type of substrate and the purity of the exogenous phosphatase, as mentioned above; the possibility of detecting alkaline nucleases such as plant exonuclease by the use of alkaline phosphatase in a reaction medium buffered at high pH;[25,32] and the use of cerium rather than lead ions as a capturing agent.[33,34] More general aspects of enzyme histochemistry, such as the choice of fixative and its effects on tissue preservation and enzyme activity, are discussed elsewhere in this handbook and by Sexton and Hall.[35]

Metal Capture Method for Detecting the 3′-Nucleotidase Activity of Plant Nuclease I
Principle[36]

As mentioned above, plant nuclease I has a 3′-nucleotidase activity; in addition to its action on phosphodiester bonds in nucleic acids, the enzyme splits off phosphate from substrates such as adenosine-3′-monophosphate (3′-AMP). It should be possible in principle to detect the nucleotidase activity cytochemically by adapting the metal capture methods proposed for 5′-nucleotidase. Lead[24] or cerium[37] ions at pH 7.3 to 8.0 are suitable capture agents for light or electron microscopy. 3′-AMP replaces 5′-AMP as substrate. Unspecific phosphatase substrates (such as β-glycerophosphate) and 5′-AMP can be used in control experiments.

Discussion

The method is speculative at present. A first requirement is that plant nuclease I is not inactivated by aldehyde fixatives. Nuclease I from *Tradescantia* probably meets this requirement; the purified enzyme was still about 80% active on DNA after treatment with 3% glutaraldehyde for an hour at 0°C.[38] Secondly, one must find a suitable pH for the reaction medium. Nuclease I acts on 3′-AMP over a very broad range of pH,[39] and lead and cerium ions can be used as capture agents up to about pH 8.0.[24,34] The main problem, then, is interference from other enzymes.

As Wilson[2] points out, three enzymes can split phosphate from 3′-AMP: nuclease I, unspecific phosphatase, and specific 3′-nucleotidase. The evidence for a completely specific 3′-nucleotidase is weak, but a 5′(3′)-nucleotidase, with high specificity for both 5′- and 3′-mononucleotides, has been isolated from wheat seedlings.[40,41] The enzyme is assayed at pH

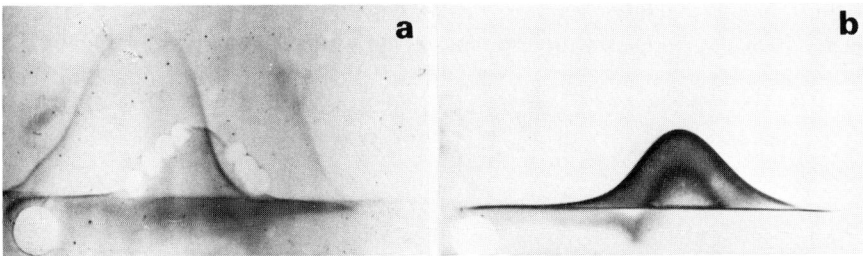

FIGURE 4. Crossed immunoelectrophoresis in agarose gels of a nuclease I preparation against antiserum. Gel (a) was stained for protein with Coomassie Brilliant Blue; gel (b) was stained histochemically for 3'-nucleotidase activity by a lead precipitation method. Gel (a) shows excision of the nuclease-nucleotidase immunoprecipitate used for a second round of immunizations. (From Smyth, C. J. and Clapham, D. H., *Hereditas*, 96, 69, 1982. With permission.)

5.0[40] and the author does not mention activity at neutral pH. It is inhibited by several ribosides. At pH 7.5 to 8.0, interference from this enzyme and from unspecific phosphatase should be minimal, so this is the best range for cytochemical tests for nuclease I.

Early attempts to demonstrate 3'-nucleotidase activity at neutral pH cytochemically were negative,[36] but recently I have had more promising results.[38] The matter seems worth further investigation. We had no problem in demonstrating the 3'-nucleotidase activity of nuclease I immunoprecipitates in agarose gels[39] (Figure 4) when we used the following medium:

Adenosine 3'-monophosphate (Sigma)	4 mg
Dissolved in distilled water	4.4 mℓ
0.2 M Tris-acetate buffer, pH 5.5	5 mℓ
2% Lead nitrate (merck)	0.6 mℓ

and incubated the gels at 37°C for 30 to 60 min.

Dye-Coupling Methods for Phosphodiesterase
Principle

Exonucleases that release 5'-mononucleotides from nucleic acids (e.g., phosphodiesterase I from animal tissues, and plant exonuclease) can split α-naphthyl thymidine 5'-phosphate and α-naphthol AS-BI thymidine 5'-phosphate.[42-45] The α-naphthol released can be coupled with a suitable dye added to the reaction medium (e.g., Fast Red TR) to yield an insoluble product marking the site of the reaction. The reaction medium must be buffered at high pH (9.2), because at neutral and acid pH other enzymes, nucleotide pyrophosphatase and cyclic phosphodiesterase, act on the substrate.[45,46] The method is applicable only to light microscopy, since the dye product is not electron dense, even after osmication. But the "indigogenic" methods for the demonstration of phosphodiesterase in ultrathin sections of animal material[44] may be adaptable to plants.

Example

In the only botanical study known to me, shoots of *Tricitum* sp. and root tips of *Vicia faba* were examined.[45] Unfixed frozen sections or frozen sections of material fixed in Baker's formal-calcium for 18 hr at 4°C were incubated at 37°C for periods up to 1 hr in the following reaction medium:

50 mM Thymidine 5'-monophosphate, α-naphthyl ester (Sigma Ltd.) or 50 mM of the α-naphthyl AS-BI ester, synthesized in the laboratory	0.1 vol
0.4 M tris-HCl buffer, pH 9.2	0.25 vol
Fast Red TR salt, final concentration 2 mg/mℓ	
Distilled water to make 1 vol	

Controls included sections inactivated by 15 min exposure to water at 90°C. The reaction product was in the cytoplasm, but the authors do not give details. (Dr. Sierakowska has recently told me that the enzyme tends to diffuse out of the sections during incubation. She suggests the possibility of using a substrate film method ([see "Substrate Film Method"] with the same reagents).

IMMUNOCYTOCHEMISTRY OF PLANT NUCLEASES

Production of Antibodies

The principles and practice of immunocytochemistry are discussed elsewhere in this handbook, so only a few general points will be taken up here. The first step in the procedure, the production of specific antibodies to the nuclease under study, can appear formidable, as there is usually no rich source of the enzyme. A well-known supplier of immunochemicals states (1983 catalog) that a conventional antiserum can be produced for a customer on receipt of 30 mg (!) of antigen, to be injected into five or six rabbits. Perhaps the firm needs no more customers. But from recent experience, it is realistic for all but the poorest immunogens to aim to raise precipitating antibodies by injecting one to two rabbits with as little as 5 to 50 μg antigen/rabbit. Further, there are tricks for obtaining suitably pure antigen for the production of conventional antisera, even if the initial enzyme preparation is not completely pure; and to raise monoclonal antibodies (usually in mice), there is no need, or even special advantage, in a 100% pure antigen for the injections.

To illustrate, we obtained an antiserum to a nuclease I preparation from *Tradescantia* that showed three immunoprecipitates with the antigen preparation in crossed immunoelectrophoresis gels.[47] (Our nuclease preparation was not strictly pure.) The nuclease precipitate was recognized by its 3'-nucleotidase activity when stained histochemically;[39] we cut out this precipitate from a total of eight gels with a cork borer (Figure 4), suspended the agarose discs in salt solution with Freund's adjuvant, and immunized two rabbits with it.[47] One of the rabbits produced a monospecific antiserum to the nuclease, responding to a few micrograms of antigen. Even submicrogram quantities of certain serum proteins can prime rabbits effectively when injected as an immunoprecipitate.[48] Bands of protein separated by electrophoresis in analytical polyacrylamide gels — even if the gels contained sodium dodecylsulfate — can also be cut out and used successfully as immunogens.[49] There is no need for milligram quantities of 100% pure antigen.

Monoclonal antibodies to plant nucleases will doubtless be used in the future. Their production and characterization are discussed elsewhere in the handbook. (It is as important to check on the specificity of monoclonal as of polyclonal antibodies, despite occasional statements to the contrary, as clones can be mixed.)

Examples of Nuclease Immunocytochemistry

Two laboratories have localized plant nucleases by immunocytochemistry. Baumgartner and Matile[50] studied an RNase in senescing petals of *Ipomoea*. They fixed pieces of petal in formaldehyde-acetic-alcohol or in 1.5% glutaraldehyde, embedded in paraffin, and cut into 7- to 10-μm sections. They located the RNase in the vacuoles by an indirect immunofluorescence method. The comprehensive controls included tests of antibody specificity by immunoelectrophoresis and a check that the immunofluorescence was absent from the vacuoles if the antiserum to RNase was first absorbed with pure antigen. The authors comment on the considerable autofluorescence of control sections not treated with immunochemicals. (This is a general feature of material fixed in glutaraldehyde, unless it is subsequently treated with sodium borohydride, glycine, or lysine to reduce free aldehyde groups.)[51] A noteworthy feature is that the antiserum to RNase was applied *undiluted* to the sections, instead of being diluted at least 1:10, but this apparently did not lead to excessive background staining (Figure 5).

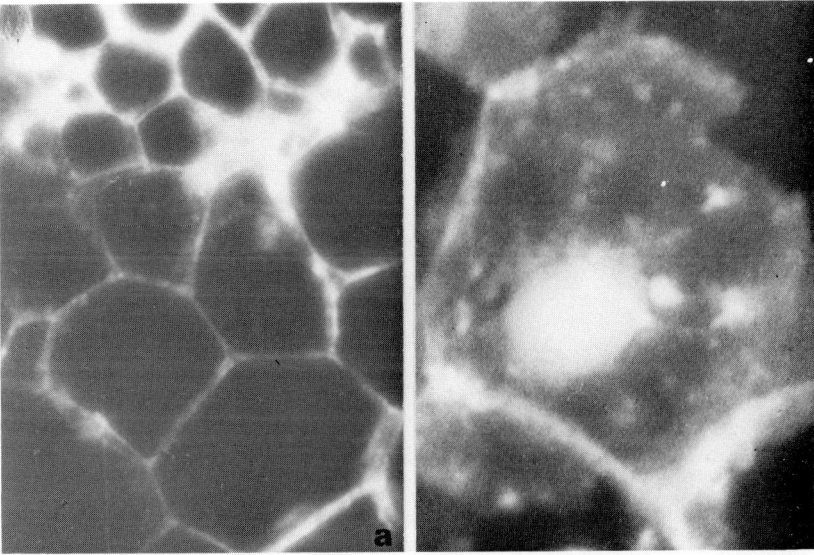

FIGURE 5. Immunocytochemical localization of an RNase in senescing *Ipomoea* petals. (a) Control section showing autofluorescence of cell wall (magnification × 100); (b) RNase-specific fluorescence of cells with preserved vacuolar contents (magnification × 2100). (From Baumgartner, B. and Matile, P., *Biochem. Physiol. Pflanzen*, 170, 279, 1976. With permission.)

The localization of an RNase in vacuoles is in good agreement with recent biochemical studies of vacuoles isolated from protoplasts[13,14] and supports the concept of the vacuole as analogous to the lysosomes of animal cells. Unfortunately, the authors[50,52,53] do not describe the properties of the *Ipomoea* RNase in sufficient detail to classify it under Wilson's scheme.

Our laboratory has localized nuclease I from *Tradescantia* by a number of immunocytochemical methods. A first paper described the purification and properties of the enzyme, enabling it to be classified as a plant nuclease I despite one or two probably minor peculiarities.[39] A monospecific antiserum was then produced, as discussed above, and characterized by crossed immunoelectrophoresis using intermediate gels.[47] Subsequently, the nuclease was localized by indirect immunofluorescence and immunoperoxidase methods,[36] immunochemicals being applied to ester or polyester wax sections of root tips that had been freeze substituted in acetone (Figure 6). The label was seen in numerous cytoplasmic particles that were provisionally identified as mitochondria owing to their size, shape, number, and general distribution in several kinds of cells. Nuclei, cell walls, vacuoles, and ground cytoplasm were considered to show at most background levels of staining. Controls included the replacement of the antinuclease solution in the staining schedule by rabbit IgG to human serum protein at the same concentration of immunoglobulins. Epidermal, mature, vascular and outer root cap cells tended to stain unspecifically.

Recently, we have extended the studies to fixed material for both light and electron microscopy.[54] For light microscopy, we fixed small pieces of root tip or leaf in buffered 3% glutaraldehyde or 4% formaldehyde, 0.5% glutaraldehyde, for 2 $1/2$ hr or overnight at 0 to 4°C. The material was later embedded in butyl methacrylate at low temperature, in order to minimize denaturation of antigen.[55]

The procedure for low temperature embedding was as follows. The tissue pieces were washed overnight in several changes of phosphate-buffered saline (PBS) at 4°C, dehydrated through an acetone series at 0°C, and allowed to cool to −20°C in pure acetone dried with anhydrous sodium sulfate. After 2 to 3 hr the tissue pieces were transferred to a mixture of

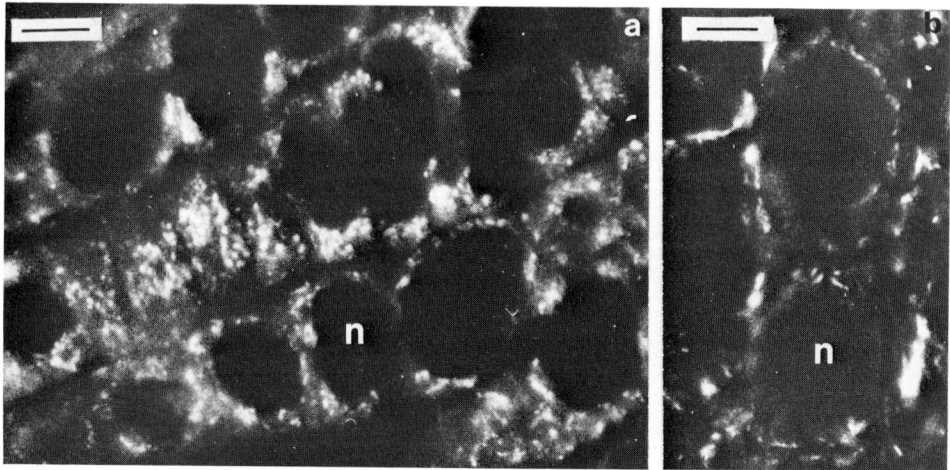

FIGURE 6. Immunofluorescent tracing of *Tradescantia* nuclease I to cytoplasmic particles; 4 μm ester wax sections of freeze-substituted root tips. (a) Close to cap junction. Antinuclease applied at 0.58 mg Ig per milliliter. Perhaps some overstaining of ground cytoplasm, but nuclei unstained; (b) cortex about 45 cells from cap junction. Antinuclease applied at 0.29 mg Ig per milliliter. Thread mitochondria or proplastids are stained, as well as rods and granules, in contrast with the ground cytoplasm. The out-of-focus fluorescent images appear greatly enlarged. Bar = 10 μm. (From Clapham, D. *Hereditas*, 96, 77, 1982. With permission.)

equal parts of acetone and butyl methacrylate containing the initiator benzoin methylether at 5 mg/mℓ. (The methacrylate had previously been treated to remove inhibitors by shaking it successively with two or three changes of 2% Na_2CO_3 and with distilled water in a separating funnel, and then drying with Na_2SO_4.) After 2 to 3 hr the tissue pieces were transferred to fresh initiated methacrylate and left to infiltrate at −20°C for a week or more with two changes of medium. Finally, the pieces were put in 8-mm Taab polythene capsules containing fresh medium and placed over dry ice in a Thermos® and left to polymerize about 15 cm from a CAMAG fluorescent tube emitting UV light at 350 nm; the temperature in the neighborhood of the capsules was about −30°C.

Sections, 3 to 4 μm, were cut dry with a glass knife on a conventional microtome and flattened and attached to slides by floating out on a drop of water at about 20°C. The knives were broken from glass strips on an LKB Histo Knifemaker® (LKB Ltd, Stockholm). The plastic was removed from the sections by immersing the slides in xylene (two changes of 1 min); the slides were then taken through acetone to PBS.

Next, sections were treated with protease (Sigma type VI) to "unmask" antigen.[56] Drops of 0.06% protease in 0.5 M tris-HCl, pH 7.5, were left on the sections for 10 min at about 20°C, after which the slides were thoroughly rinsed in several changes of PBS for 5 to 10 min.

Antinuclease solution was now applied to the sections, diluted in PBS containing 20% normal swine serum to prevent unspecific staining. The antinuclease preparation originally contained about 30 mg IgG per milliliter and was applied to the sections at a dilution of 1:130 for 30 min at about 20°C. The slides were then rinsed in PBS and the second antibodies applied; these consisted of swine antirabbit IgG coupled with peroxidase (Dako Ltd.) diluted 1:100 in PBS containing 20% normal swime serum

Slides were subsequently processed through diaminobenzidine in a standard way and mounted in Canada balsam. They were examined in bright field and phase contrast, the latter being surprisingly effective for showing up unstained regions of the cells in good contrast with the stained particles (Figure 7a). Controls were as described previously and, in addition, it was found that staining was absent from sections that were treated with

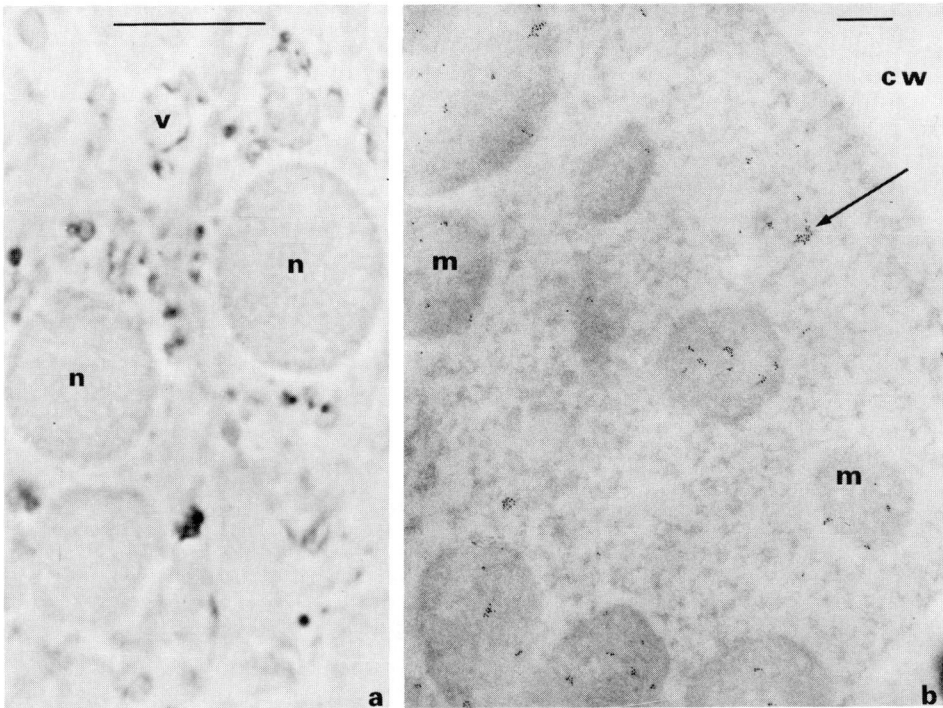

FIGURE 7. Immunocytochemistry of nuclease I. (a) Immunoperoxidase localization of nuclease in cytoplasmic particles (dark staining), presumably mitochondria, in *Tradescantia* root-tip cells. Nuclei (n) and vacuoles (v) are unstained. Phase micrograph. Bar = 10 μm; (b) immunogold localization of nuclease in mitochondria (m) and cytosol of a mesophyll cell of a *Pisum* leaf infected with red clover mottle virus. Gold particles (arrows) are absent from cell wall (cw). Bar = 200 nm. (From Clapham, D. H. and Tomenius, K., unpublished.)

antinuclease that had been absorbed with purified antigen for 2 days at 4°C. (The antigen was not, however, 100% pure.)

The results (Figure 7a) confirm the work with unfixed, freeze-substituted root tips — the nuclease appears to be localized in the mitochondria, as far as can be judged by light microscopy.

To extend the work to the level of the electron microscope,[54] we embedded the material at low temperature in the newly developed methacrylate-acrylate medium, Lowicryl K4M[57] (Juniper Ultra Micro, Stockholm), and, as second antibodies, we applied swine antirabbit IgG coupled with colloidal gold.[54,55,58] (The particulate gold label is easier to see in electron micrographs than the diffuse label of the immunoperoxidase system.) The procedure for embedding was essentially as just described for butyl methacrylate. Methods for embedding at low temperature, and for coupling colloidal gold to immunoglobulins or to protein A, are under development and are discussed elsewhere in this handbook; here I shall just cite some recent botanical papers.[55,59-62] We find that affinity purification[51,63] of the second antibodies (swine antirabbit IgG, Dakopatts Ltd.) on normal rabbit IgG (Dakopatts), before coupling with colloidal gold of nominal diameter 5 nm, results in lower background staining.

Some results are shown in Figure 7b (viral-infected pea leaf mesophyll; the antiserum to *Tradescantia* nuclease I reacts strongly with nuclease I from other flowering plants and even with P1 nuclease from *Penicillium*).[38] Ultrathin sections were treated with 5% bovine serum albumin in PBS for 15 min, to reduce unspecific staining, then with antinuclease at a concentration of 30 μg Ig per milliliter in albumin-PBS overnight at 6 to 8°C. The grids were then rinsed and treated with gold-labeled second antibodies diluted in albumin-PBS

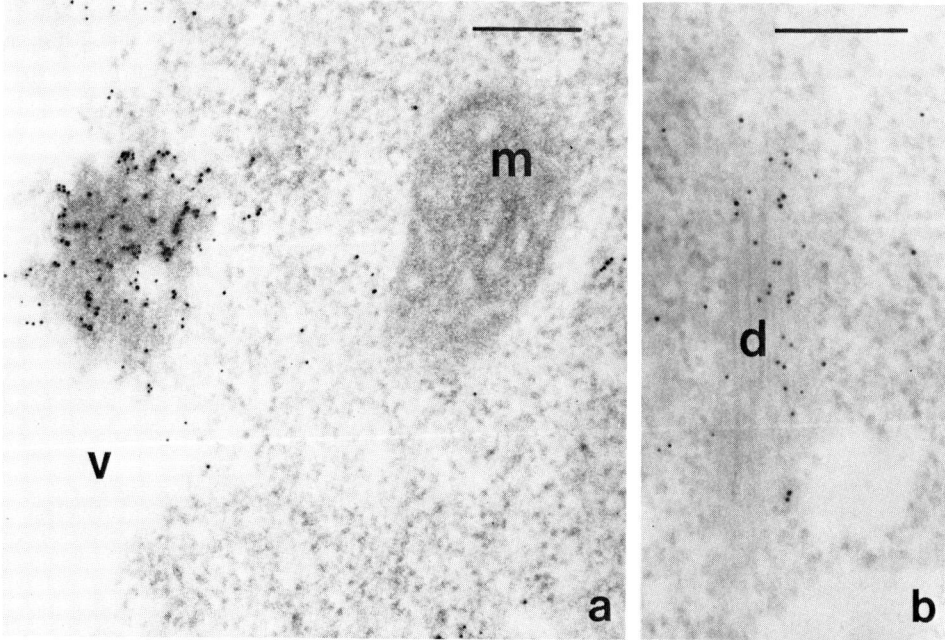

FIGURE 8. Immunogold localization of nuclease I in *Tradescantia* root tips using antibodies affinity-purified on P1 nuclease from *Penicillium*. (a) Unlabeled mitochondrion (m) in ground cytoplasm and heavily labeled structure, perhaps a mitochondrion, in the vacuole (v). (b) Dictyosome (d), asymmetrically labeled. Bar = 200 nm. (From Clapham, D. H. and Tomenius, K., unpublished.)

for 15 min, rinsed again, treated with uranyl acetate, and examined. Staining was carried out at 20 to 23°C if not otherwise stated.

As regards controls, label was virtually absent from sections in which the first antibody was omitted or replaced with appropriately diluted, unrelated antiserum, or when the first antibodies had been absorbed with antigen.

Very recently we have obtained better structure by infiltrating the material with Lowicryl K4M at +5° C, and clearer labeling patterns by using antinuclease antibodies that have been affinity-purified against P1 nuclease, and by including Tween 20 and bovine serum albumin in the staining and wash solutions[54] (Figure 8). The distribution of label is then much as expected for acid hydrolases. Dictyosomes are consistently and strongly labeled. Structures presumed to be mitochondria and proplastids are strongly labeled when they are in or close to small vacuoles and apparently being digested; these, together with strongly labeled crystals in the vacuoles and perhaps some of the dictyosomes are what must show up in preparations for light microscopy. There are labeled protuberances from the tonoplasts of small vacuoles and from the plasma membrane, the latter being labeled particularly in the neighborhood of plasmodesmata. A small fraction of the endoplasmic reticulum is labeled. Structures that are not labeled include cell walls, nuclei, chloroplasts, peroxisomes, vacuolar sap and the tonoplasts of large vacuoles.

Attractive features of immunogold cytochemistry are the ease with which the gold label can be recognized and quantitated in ultrathin sections. The label tends to appear in small clumps, however, the clumping being more or less pronounced according to the method of coupling the colloidal gold to the immunoglobulins.[64,65] Protein A-gold does not form clumps, but is more liable to bind unspecifically to plant tissues.

Can the immunocytochemistry be trusted? Hazards are (1) false-negative results, i.e., the antigen appears to be absent from one or more sites because it was inactivated or "masked" during fixation and embedding, or, alternatively, was lost from inadequately fixed sections; and (2) false-positive results, owing to the presence of antibodies of the wrong specificity in the antinuclease solution, or to redistribution of poorly fixed antigen.

Fixation is a critical step here, as always in cytochemistry. The cells are well fixed apart from membranes, which mostly appear in negative contrast. This is as expected, since we used fixatives (4% paraformaldehyde, 0.5% glutaraldehyde, or 3% glutaraldehyde) that fix protein well while failing to react with lipids unless they contain $-NH_2$ groups. So fixation was probably adequate to keep the antigen in place; nor is the *Tradescantia* nuclease particularly labile to aldehyde fixatives, since the antigenicity of the isolated enzyme bound to a microtiter plate is reduced only 25% after overnight exposure to 3% glutaraldehyde at 4°C.[38,66] Low temperature embedding reduces further inactivation at this stage.

As regards the antibodies, the antinuclease is monospecific, as judged by crossed immunoelectrophoresis,[47] and was applied at reasonable dilution for immunocytochemistry. Nuclease I, however, is a glycoprotein,[39] and antisera to one glycoprotein sometimes cross react with the carbohydrate fraction of unrelated glycoproteins.[67,68] But our present antinuclease solution does not react in crossed immunoelectrophoresis with seven to ten other glycoproteins copurifying with nuclease I, although an earlier polyspecific antiserum did so;[47] and the affinity-purified antibodies do not react with horse radish peroxidase or potato acid phosphatase bound to microtiter plates. It is unlikely, therefore, that such cross reactions are affecting our conclusions. But we shall take the matter further with monoclonal antibodies directed against the protein component of nuclease I.

Recently, some secretory proteins have been demonstrated by a protein A-gold method applied to ultrathin sections of material fixed in glutaraldehyde, osmicated, and embedded in Epon,[69] and by an immunogold method applied to osmicated material embedded in the acrylic resin LR White (London Resin Company, Basingstoke, U.K.).[70] Antigen in the sections is reactivated by treatment with saturated sodium mateperiodate, sometimes followed by decinormal HCl before immunocytochemical staining. Structural preservation is considerably superior to that obtained with material fixed in glutaraldehyde alone and embedded in Lowicryl.

CONCLUSIONS

A botanist wishing to locate a nuclease is advised to develop an immunocytochemical method. None of the various proposals for enzyme cytochemistry is fully satisfactory, because even in the best cases, one does not know which nuclease is being studied. (Plant physiologists can talk of "RNase activity" when they would not dream of saying "carbohydrate-splitting activity".) Yet, the enzyme cytochemical methods are worth developing. For example, a metal capture method reasonably specific for nuclease I might be able, unlike antibodies, to distinguish active from inactive forms of the enzyme.

Nuclease immunocytochemistry is an attractive field. The effort needed to purify the enzyme of interest and raise antibodies can be daunting. But the enzyme has, in any case, to be purified if it is to be defined, owing to the multiplicity of nucleases, and quantitative studies of the individual enzymes in tissues and organs are probably feasible only by immunoassay. Antibodies are useful outside cytochemistry.

In the future, nucleases present in plant cells only in tiny amounts may well be produced in milligram quantities in bacteria,[71] enabling the production of antibodies, conventional or monoclonal. This will open the way to cytochemical studies by immunogold methods of nucleases involved in the synthesis, repair, and recombination of DNA, and in the splicing and other processing of RNA.

ACKNOWLEDGMENTS

I wish to thank Dr. Cyril Smyth and Dr. Karin Tomenius for their collaboration, Prof. Gunnar Östergren, Dr. Herman Amneus, Dr. Christer Andersson, and Dr. Lena Clapham for commenting on the manuscript, and the Swedish Council for Forestry and Agricultural Research for financial support.

REFERENCES

1. **Kornberg, A.,** *DNA Replication,* Freeman, San Francisco, 1980, chap. 10.
2. **Wilson, C. M.,** Plant nucleases, *Ann. Rev. Plant Physiol.,* 26, 187, 1975.
3. **Farkas, G. L.,** Ribonucleases and ribonucleic acid breakdown, in *Encyclopedia of Plant Physiology, New Series,* Vol. 14B, Parthier, B. and Boulter, D., Eds., Springer-Verlag, Berlin, 1982, 224.
4. **Linn, S.,** Nucleases involved in DNA repair, in *Nucleases,* Linn, S. M. and Roberts, R. J., Eds., Cold Spring Harbor Laboratory, Cold Spring Harbor, N.Y., 1982, 59.
5. **Linn, S.,** The deoxyribonucleases of *Escherichia coli,* in *Nucleases,* Linn, S. M. and Roberts, R. J., Eds., Cold Spring Harbor Laboratory, Cold Spring Harbor, N.Y., 1982, 306.
6. **Howell, S. H. and Stern, H.,** The appearance of DNA breakage and repair activities in the synchronous meiotic cycle of *Lilium, J. Mol. Biol.,* 55, 357, 1971.
7. **Thibodeau, L. and Verly, W. G.,** Endonucleases for apurinic sites in plants, *FEBS Lett.,* 69, 183, 1976.
8. **Szopa, J., Böcher, M., and Wagner, K. G.,** Purification and properties of an endonuclease from nuclei of tobacco cell cultures, *Z. Pflanzenphysiol.,* 109, 291, 1983.
9. **Szopa, J., Szmidzinski, R., and Wagner, K. G.,** Purification and properties of a manganese-dependent endonuclease from tobacco cell cultures, *Z. Pflanzenphysiol.,* 111, 341, 1983.
10. **Kroeker, W. D. and Fairley, J. L.,** The limited cleavage of bihelical deoxyribonucleic acid by wheat seedling nuclease, *J. Biol. Chem.,* 250, 3773, 1975.
11. **Kroeker, W. D., Kowalski, D., and Laskowski, M.,** Mung bean nuclease I. Terminally directed hydrolysis of native DNA, *Biochem. (Wash.),* 15, 4463, 1976.
12. **Linn, S. M. and Roberts, R. J., Eds.,** *Nucleases,* Cold Spring Harbor Laboratory, Cold Spring Harbor, N.Y., 1982.
13. **Nishimura, M. and Beevers, H.,** Hydrolases in vacuoles from castor bean endosperm, *Plant Physiol.,* 62, 44, 1978.
14. **Boller, T. and Kende, H.,** Hydrolytic enzymes in the central vacuole of plant cells, *Plant Physiol.,* 63, 1123, 1979.
15. **Daoust, R.,** Localization of deoxyribonuclease activity in tissue sections: a new approach to the histochemistry of enzymes, *Exp. Cell Res.,* 12, 203, 1957.
16. **Daoust, R.,** Localization of deoxyribonuclease activity by the substrate film method, in *General Cytochemical Methods,* Vol. 2, Danielli, J. F., Ed., Academic Press, New York, 1961, 153.
17. **Sierakowska, H. and Shugar, D.,** Gross histochemical localization of tissue nuclease enzymes, *Acta Biochim. Pol.,* 8, 427, 1961.
18. **Daoust, R.,** Histochemical localization of enzyme activities by substrate film methods, in *International Review of Cytology,* Vol. 18, Bourne, G. H. and Danielli, J. F., Eds., Academic Press, New York, 1965, 191.
19. **Daoust, R.,** Modified procedure for the histochemical localization of ribonuclease activity by the substrate film method, *J. Histochem. Cytochem.,* 14, 254, 1966.
20. **Daoust, R.,** Histochemical comparison of focal losses of RNase and ATPase activities in preneoplastic rat livers, *J. Histochem. Cytochem.,* 27, 653, 1979.
21. **Knox, R. B. and Heslop-Harrison, J.,** Pollen-wall proteins: localization and enzymic activity, *J. Cell Sci.,* 6, 1, 1970.
22. **Aronson, J., Hempelmann, L. H., and Okada, S.,** Preliminary studies on the histological demonstration of desoxyribonuclease II by adaptation of the Gomori acid phosphatase method, *J. Histochem. Cytochem.,* 6, 255, 1958.
23. **Vorbrodt, A.,** Histochemical studies on the intracellular localization of acid deoxyribonuclease, *J. Histochem. Cytochem.,* 9, 647, 1961.
24. **Gomori, G.,** *Microscopic Histochemistry,* Chicago University, Chicago, 1952.

25. **Taper, H. S.,** Evaluation of the validity of the histochemical lead nitrate technique for alkaline and acid deoxyribonuclease, *J. Histochem. Cytochem.,* 27, 1483, 1979.
26. **Vaughn, K. C.,** Cytochemical demonstration of DNase in pollen, *Histochemistry,* 73, 487, 1982.
27. **Vaughn, K. C.,** letter to author, 1983.
28. **Clapham, D.,** Contamination of commercial acid phosphatases with nuclease, *Protoplasma,* 107, 109, 1981.
29. **Coulomb, P.,** Localisation de la désoxyribonucléase acide dans le méristème radiculaire de la courge (*Cucurbita pepo* L. Cucurbitacée), *C. R. Acad. Sci. Paris,* 269, 1514, 1969.
30. **Gahan, P. B., Perry, I. J., Stroun, M., and Anker, P.,** Effect of exogenous DNA on acid deoxyribonuclease activity in intact roots of *Vicia faba* L., *Ann. Bot.,* 38, 701, 1974.
31. **Laskowski, M.,** Paper chromatography and characterization of oligodeoxyribonucleotides, in *Methods in Enzymology,* Vol. 12, Grossman, L. and Moldave, K., Eds., Academic Press, New York, 1967, 281.
32. **Zotikov, L. and Bernhard, W.,** Localisation au microscope électronique de l'activité de certaines nucléases dans des coupes à congélation ultrafines, *J. Ultrastruct. Res.,* 30, 642, 1970.
33. **Veenhuis, M., Van Dijken, J. P., and Harder, W.,** A new method for the cytochemical demonstration of phosphatase activities in yeasts based on the use of cerous ions, *FEMS Microbiol. Lett.,* 9, 285, 1980.
34. **Robinson, J. M. and Karnovsky, M. J.,** Ultrastructural localization of several phosphatases with cerium, *J. Histochem. Cytochem.,* 31, 1197, 1983.
35. **Sexton, R. and Hall, J. L.,** Enzyme cytochemistry, in *Electron Microscopy and Cytochemistry of Plant Cells,* Hall, J. L., Ed., Elsevier/North-Holland, Amsterdam, 1978, 63.
36. **Clapham, D.,** Immunocytochemistry of nuclease 1 of *Tradescantia, Hereditas,* 96, 77, 1982.
37. **Robinson, J. M. and Karnovsky, M. J.,** Ultrastructural localization of 5'-nucleotidase in guinea pig neutrophils based upon the use of cerium as capturing agent, *J. Histochem. Cytochem.,* 31, 1190, 1983.
38. **Clapham, D. H.,** unpublished observations.
39. **Clapham, D.,** Properties of plant nuclease 1 purified from *Tradescantia* leaves by binding to Concanavalin A-Sepharose, *Hereditas,* 93, 137, 1980.
40. **Polya, G. M.,** Regulation of a plant 5'(3')-ribonucleotide phosphohydrolase by cyclic nucleotides and pyrimidine, purine and cytokinin ribosides, *Proc. Natl. Acad. Sci. U.S.A.,* 71, 1299, 1974.
41. **Fritzson, P.,** Regulation of nucleotidase activities in animal tissues, *Adv. Enzyme Regul.,* 16, 43, 1978.
42. **Sierakowska, H., Szemplinska, H., and Shugar, D.,** Intracellular localization of phosphodiesterase by a cytochemical method, *Acta Biochim. Polon.,* 10, 399, 1963.
43. **Shugar, D. and Sierakowska, H.,** Mammalian nucleolytic enzymes and their localization, *Prog. Nucl. Acids. Res.,* 7, 369, 1967.
44. **Tsou, K. C.,** 5'-Nucleotide phosphodiesterase, in *Electron Microscopy of Enzymes,* Vol. 4, Hayat, M. A., Ed., Van Nostrand Reinhold, New York, 1975, 140.
45. **Sierakowska, H., Gahan, P. B., and Dawson, A. L.,** The cytochemical localization of nucleotide pyrophosphatase activity in plant tissues using naphthyl esters of thymidine-5'-phosphate, *Histochem. J.,* 10, 679, 1978.
46. **Bartkiewicz, M. and Sierakowska, H.,** Histochemical localization of nucleotide pyrophosphatase and cyclic nucleotide phosphodiesterase in seeds and shoots of *Triticum, Planta,* 155, 204, 1982.
47. **Smyth, C. J. and Clapham, D. H.,** Uso of immunoprecipitates to obtain monospecific immunoglobulins to nuclease 1 of *Tradescantia, Hereditas,* 96, 69, 1982.
48. **Crowle, A. J., Revis, G. J., and Jarrett, K.,** Preparatory electroimmunodiffusion for making precipitins to selected native antigens, *Immunol. Commun.,* 1, 325, 1972.
49. **Lazarides, E.,** Antibody production and immunofluorescent characterization of actin and contractile proteins, in *Methods in Cell Biology,* Vol. 24, Wilson, L., Ed., Academic Press, New York, 1982, 313.
50. **Baumgartner, B. and Matile, P.,** Immunocytochemical localization of acid ribonuclease in morning glory flower tissue, *Biochem. Physiol. Pflanzen,* 170, 279, 1976.
51. **Osborn, M. and Weber, K.,** Immunofluorescence and immunocytochemical procedures with affinity purified antibodies: tubulin-containing structures, in *Methods in Cell Biology,* Vol. 24, Wilson, L., Ed., Academic Press, New York, 1982, 97.
52. **Baumgartner, B., Kende, H., and Matile, P.,** Ribonuclease in senescing morning glory. Purification and demonstration of *de novo* synthesis, *Plant Physiol.,* 55, 734, 1975.
53. **Baumgartner, B. and Matile, P.,** Isoenzymes of RNase in senescing morning glory petals, *Z. Pflanzenphysiol.,* 82, 371, 1977.
54. **Clapham, D. H. and Tomenius, K.,** Immunocytochemistry of plant nuclease I, submitted for publication.
55. **Tomenius, K., Clapham, D., and Oxelfelt, P.,** Localization by immunogold cytochemistry of viral antigen in sections of plant cells infected with red clover mottle virus, *J.Gen. Virol.,* 64, 2669, 1983.
56. **Denk, H., Radaszkiweicz, Y., and Weirich, E.,** Pronase pretreatment of tissue sections enhances sensitivity of unlabelled antibody-enzyme (PAP) technique, *J. Immunol. Meth.,* 15, 163, 1977.

57. **Roth, J., Bendayan, M., Carlemalm, E., Villiger, W., and Garavito, M.,** Enhancement of structural preservation and immunocytochemical staining in low temperature embedded pancreatic tissue, *J. Histochem. Cytochem.*, 29, 663, 1981.
58. **De Mey, J., Moeremans, M., Geuens, G., Nuydens, R., and De Brabander, M.,** High resolution light and electron microscopic localization of tubulin with the IGS (immuno gold staining) method, *Cell Biol. Int. Rep.*, 5, 889, 1981.
59. **Craig, S. and Millerd, A.,** Pea seed storage proteins — immunocytochemical localization with protein A-gold by electron microscopy, *Protoplasma*, 105, 333, 1981.
60. **Craig, S. and Goodchild, D. J.,** Post-embedding immunolabeling. Some effects of tissue preparation on the antigenicity of plant proteins, *Eur. J. Cell Biol.*, 28, 251, 1982.
61. **Herman, E. M. and Shannon, L. M.,** Immunocytochemical evidence for the involvement of Golgi apparatus in the deposition of seed lectin of *Bauhinia purpurea (Leguminosae)*, *Protoplasma*, 121, 163, 1984.
62. **Lin, N. S. and Langenberg, W. G.,** Immunohistochemical localization of barley stripe mosaic virions in infected wheat cells, *J. Ultrastruct. Res.*, 84, 16, 1983.
63. **Fuller, G. M., Brinkley, B. R., and Boughter, M. J.,** Immunofluorescence of mitotic spindles by using monospecific antibody against bovine brain tubulin, *Science*, 187, 948, 1975.
64. **De Mey, J.,** Colloidal gold probes in immunocytochemistry, in *Immunocytochemistry*, Polak, J. M. and Van Noorden, S., Eds., Wright, PSG, Bristol, U.K., 1983, 82.
65. **Slot, J. W. and Geuze, H. J.,** Gold markers for single and double immunolabelling of ultrathin cryosections, in *Immunolabelling for Electron Microscopy*, Polak, J. M. and Varndell, I. M., Eds., Elsevier, Amsterdam, 1984, 129.
66. **Kennett, R. H.,** Enzyme-linked antibody assay with cels attached to polyvinyl chloride plates, in *Monoclonal Antibodies*, Kennett, R. H., McKearn, T. J., and Bechtol, K. B., Eds., Plenum Press, New York, 1980, 376.
67. **Howlett, B. J. and Clarke, A. E.,** Isolation and partial characterization of two antigenic glycoproteins from rye-grass *(Lolium perenne)* pollen, *Biochem. J.*, 197, 695, 1981.
68. **Howlett, B. J. and Clarke, A. E.,** Role of carbohydrate as an antigenic determinant of a glycoprotein from rye-grass *(Lolium perenne)* pollen, *Biochem. J.*, 197, 707, 1981.
69. **Bendayan, M. and Zollinger, M.,** Ultrastructural localization of antigentic sites on osmium-fixed tissues applying the protein A-gold technique, *J. Histochem. Cytochem.*, 31, 101, 1983.
70. **Craig, S. and Miller, C.,** LR white resin and improved on-grid immunogold detection of vicilin, a pea seed storage protein, *Cell Biol. Int. Rep.*, 8, 879, 1984.
71. **Alberts, B., Bray, D., Lewis, J., Raff, M., Roberts, K., and Watson, J. D.,** *Molecular Biology of the Cell*, Garland, New York, 1983, 191.

CYTOCHEMICAL LOCALIZATION OF LIPASES IN PLANT CELLS

Maria Salomé S. Pais

INTRODUCTION

Lipases in Animal Cells

The term lipases has been used for the first time by Hanriot[1] to name the enzyme responsible for the hydrolyses of monobutyrin in the serum and animal tissues such as liver, stomach, and pancreas. This enzyme was recognized biochemically in these tissues and has also been demonstrated histochemically.[2] The method proposed was based on the same principle as that used for histochemical location of phosphatase activity. Several authors have improved on the original histochemical method for demonstration of lipase activity.[3-7] The first report on the ultrastructural location of lipase in liver cells is due to Mizuhira and Kurotaki[8] who used Tween 40 as substrate. Later on, a new procedure was proposed for light microscopic visualization of lipase activity.[9] The authors have used naphtol AS nonanoate as substrate and sodium taurocholate as activator of lipase activity. Although pancreatic lipase did hydrolyze naphtylalkanoates, the blue azo dye is lipid soluble and extracted by dehydration and embedding and did not produce enough electron opacity to be used in electron microscopic cytochemistry. Having recognized the disadvantage of this method, Hanker et al.[10] have proposed a new method for location of lipases and esterases based on the reducing properties of thiolesters reacting with osmium tetroxide vapors. In the sequence of this reaction, an electron-dense nonsoluble deposit is formed. Seligman et al.[11] have developed a technique based on the formation of diazothioethers from 2-thiolnonanoylbenzanilide in dimethylacetamide and diazonium salt (Fast Blue BBN) which can be converted in electron-dense deposits after exposure to osmium tetroxide vapors. Nevertheless, the osmium deposits are not selective enough to allow a precise location of lipase activity.

Gomori's original method was improved for cytochemical location of lipases.[12,13] This method allows the formation of a very fine lead precipitate in the sites of lipase activity. The Gomori original method has often been called the "Tween method".[14] Takigami et al.[15] have compared the results on cytochemical demonstration of Tween lipases in the pancreatic cells when various amounts of glycerin, dimethylsulfoxide (DMSO), or Triton X-100 were added to the incubation medium at pH 6 to 8.

An immunocytochemical method, based on the protein A-gold technique, was developed for localization of lipase activity in rat pancreas.[16] The methods reported were only applied to locate lipase activity at pH 7.0. To the best of our knowledge, only one histochemical method has been developed to locate acid lipase in animal cells.[17] This method was based on the histochemical diazocoupling technique applied to the product of the hydrolysis of a naphtyl palmitate dispersed with Triton X-100 at acid pH.

Lipases in Plant Cells

The presence of lipases (acid, neutral, and alkaline) has been insistently recognized in biochemical studies of plant tissues, especially during germination of oil seeds.[18-22]

Histochemical localizations of lipase activity at the light microscope level have been carried out by different authors using different procedures. Furr and Mahlberg[23] have identified lipase activity using a sulfide reaction according to Chayen et al.[24] This method was based on the formation of a lead sulfide precipitate formed as a reaction product of lead nitrate and hydrogen sulfide water. An electron microscopic cytochemical procedure for acid, neutral and alkaline lipase defection during spore germination has recently been developed[25] that utilizes similar protocols.

Biochemical evidence for the presence of phospholipase A, C, and D in plant cells has long been available. The presence of phospholipase A_2 in the nonphotosynthetic chromoplast membranes has been reported.[26] The action of phospholipase D on phospholipid degradation has also been referred.[27,28] The activity of phospholipase was also correlated with the decomposition of the spherosomes membrane of soybean.[29]

A method for cytochemical location of phospholipase A and lysophospholipase was developed by Pug and Cawson[30] to locate the activity of these enzymes in the cells of *Candida albicans*. This method is based on the ability of phospholipase A to hydrolyze lecithin with the production of lysolecithin and fatty acid. The lysolecithin formed is split by lysolecithinase into glycerylphosphorylcholine and a fatty acid.[31] The enzyme activity is located by coupling lead from the incubation mixture with the released fatty acids. In these conditions, the metal deposit obtained after incubation in a mixture containing lecithin as a substrate will demonstrate the combined enzyme activities of phospholipase A and lysophospholipase.

Comparatively, only one enzyme (lysophospholipase = lysolecithinase) can be located if the substrate used is lysolecithin.

CYTOCHEMICAL PROCEDURE

Fixation

Spores or fresh tissue slices with approximately 1 mm³ are prefixed in 2.5% glutaraldehyde solution buffered with 0.1 M cacodylate, pH 7.2, at 4°C during a period of 1 hr.

Longer periods of fixation were tested and no significant inhibition was detected with fixations as long as 5 hr.

The glutaraldehyde solution was prepared immediately prior to use as follows:

25% Highly purified glutaraldehyde	1 mℓ
0.1 M cacodylate buffer, pH 7.2	9 mℓ

Incubation

One-Step Method

When this method is used, the lead ions function as capture reagent. In this case, the composition of the incubation medium is as follows:

(1) 5% Tween 80 or Tween 60 solution	1.0 mℓ
0.2 Tris buffer (pH 7.2)	2.5 mℓ
0.08 To 0.02% aqueous lead nitrate solution	1.0 mℓ
Distilled water	20.5 mℓ

The one step method was used successfully to locate lipoprotein lipase in adipose tissue.[32] Nagata[13] was not successful in locating lipase activity by using the one-step method. This author pointed out that the inability of the one-step method to locate alkaline lipase activity, in contrast with good results obtained on the location of alkaline phosphatase activity,[33] must be due to the rapid inhibition of the enzyme activity when the specimen is exposed to the lead ions simultaneously with the substrate. If the enzyme and the substrate are complexed prior to the addition of lead ions, the enzyme maintains its activity.

We have not tested the one-step method for location of lipase activity in plant material nor are there any literature reports where this method has been successfully utilized.

Two-Step Method

After fixation, tissue slices are hand sectioned or sectioned in freezing microtome into sections approximately 50 μm thick. This material is submitted to a preincubation in the following medium:

(2) 0.2 *M* Tris buffer (pH 7.2)	2.5 mℓ
Sodium fluoride	0.1 mg
Distilled water	22.5 mℓ

Spores are preincubated in the same medium immediately after washing.

Incubation Media

(3) 5% Tween 80 or Tween 60 solution	1.0 mℓ
0.2 M Tris buffer (pH 7.2)	2.5 mℓ
10% Aqueous calcium chloride	1.0 mℓ
Distilled water	20.5 mℓ
(4) Tween 80 or Tween 60 solution	1.0 mℓ
0.2 *M* Tris buffer (pH 7.2)	2.5 mℓ
10% Aqueous calcium chloride	1.0 mℓ
2.5% Aqueous sodium taurocholate	2.5 mℓ
Distilled water	18.0 mℓ
(5) 0.2 *M* Tris buffer (pH 7.2)	2.5 mℓ
10% Aqueous calcium chloride	1.0 mℓ
Distilled water	21.5 mℓ
(6) 5% Tween 80 or Tween 60 solution	1.0 mℓ
0.2 *M* Tris buffer (pH 7.2)	2.5 mℓ
10% Aqueous calcium chloride	1.0 mℓ
Quinine hydrochloride	0.099 mg
Distilled water	20.5 mℓ

Tween 80 or Tween 60 can be used as substrate. Tween 80, as an unsaturated fatty acid ester (oleic acid ester), is a liquid and for this reason it can easily be measured with a pipette.

The solution of Tween 80, when prepared by dissolving 0.5 mℓ of Tween 80 in 9.5 mℓ distilled water, can be stored in the refrigerator for several months.

Tween 60 (estearic acid ester) solution is prepared as follows: Tween 60 is heated to melting point; 0.5 mℓ of Tween 60 is dissolved in 9.5 mℓ of distilled water immediately before use.

To prepare 0.2 *M* Tris buffer (pH 7.2), dissolve 29 mg maleic acid and 30.3 mg Tris (hydroxymethyl) aminomethane in 500 mℓ of distilled water.

Acid Lipase

The incubation media are the same as those used for neutral lipase with 0.2 *M* Tris buffer pH adjusted to 5.0. The pH is adjusted with HCl 0.2 *N*.

Alkaline Lipase

Just as for acid lipase, the incubation media only differ from those of neutral lipase in the pH value (pH 9.0) of 0.2 *M* Tris buffer. The pH was adjusted with NaOH 0.2 *N*.

The final pH of all the incubation media must be verified.

According to the tissue, a suitable molarity of the fixative mixture and of the incubation media must be kept.

Incubation Temperature

Incubation is carried out at 37°C in a shaking bath at 70 strokes/min. Duration of incubation, between 1 to 18 hr, depending on the material, can be used.

As was reported by Nagata and Murata,[12] the size of the deposits of the reaction product depends on the incubation time and it increases with increase of incubation time.

With our material, the finest results were obtained with a period of 3 hr as incubation time.

Substitution with Lead Solution

Material incubated in medium 3, 4, 5, or 6 is washed for 3 min in 2% EDTA in cacodylate buffer (pH 7.2) to remove calcium according to Adams et al.[34]

To prepare EDTA dissolve 2 g EDTA in 100 mℓ cacodylate buffer.

The pH of cacodylate buffer must be adjusted with HNO_3.

After washing in 0.1 M cacodylate buffer, pH 7.2, calcium is substituted by immersion of specimens in 0.15% aqueous lead nitrate solution for 10 min at room temperature.

To prepare nitrate solution, dissolve 0.15 g lead nitrate in 100 mℓ distilled water. Distilled water must be boiled.

Postfixation

Specimens prepared according to the two-step method were submitted to a postfixation.

In this case, the tissue specimens, after immersion in lead nitrate solution, are washed twice in 0.1 M cacodylate buffer for 5 min and postfixed with 1% osmium tetroxide solution in cacodylate buffer for 1 to 2 hr at 4°C.

After dehydration by a graded acetone series, material is embedded in Araldite or Epon-Araldite. Ultrathin sections, gray to gold, must be observed without staining. Nagata[13] used sections stained with uranyl acetate followed by lead citrate.

RESULTS

One of the major problems to be solved in the cytochemical location of lipases activity is the differentiation of lipases from nonspecific esterases and nonspecific lead deposits. Some features can condition the specificity of the reaction.

Type of Substrate

Saturated (Tween 20, 40, and 60) and unsaturated fatty acid ester (Tween 80) have been used by Gomori, who has emphasized that the use of Tween 80 as substrate demonstrates the sites of true lipase activity.[2] Bruno and Marino[35] have reported that Tween 80 locates either lipase or nonspecific esterase activity. Later on, it has been demonstrated that Tween 80 is decomposed by nonspecific esterases.[36] According to Diaconita,[37] the incubation medium containing Tween 60 at pH 7.4 is not convenient to distinguish between lipase and esterase in the guinea pig lung cells. These authors suggested the use of Tween 80 at pH 8.5 for lipase location. In our experiments, we could not find any difference between reactions using Tween 60 or Tween 80. Similar conclusions have been reported for animal cells.[13]

Nonspecific Lead Depositions

Nonspecific lead depositions are, in great part, due to the concentration of lead nitrate solutions, or to the presence of Cl^-, HCO_3^-, or HSO_4^-. Reagents must, therefore, be pure enough to contain very small amounts of these anions. pH adjustments of washing buffers and of solutions used after incubation cannot be accomplished with H_2SO_4 or HCl. Distilled water must be boiled to remove CO_2 to prevent formation of HCO_3^-.

Lead nitrate concentrations must be kept below 0.25%. The best results were obtained with concentrations between 0.1 and 0.15%. In these conditions, lead deposition is very selective in our material.

It has been mentioned that the pH value of the solution can also be responsible for the nonspecific lead deposition.[13] The use of sucrose solutions must be avoided in the cytochemical detection of lipase activity.[13] The use of this sugar triggers the formation of nonspecific lead deposition in the cytosol and in the nucleus similar to that described for cytochemical location of acid phosphatase activity.[38]

Our experiments, taking into account the precautions mentioned on the presence of the

Table 1
SOME INHIBITORS AND ACCELERATORS OF LIPASE AND ESTERASE ACTIVITY

Lipase activator	Lipase inhibitor	Esterase inhibitor	Ref.
	Eserine (3.5×10^{-3} M)	Eserine (10^{-1} M) sodium arseniate	39
		(10^{-1} M)	40
		Fluoride (3 mg mℓ^{-1})	40
Sodium taurocholate (10^{-2} M)	Quinine hydrochloride (10^{-2} M)	Sodium taurocholate (10^{-2} M)	40
		Atoxyl (10^{-2} M)	40
		Di-iso-propylfluorophosphate (10^{-6} M)	41
		Trichlorphone (10^{-5} M)	42
		Diethyl p-nitrophenylphosphate (10^{-5} M)	42

referred anions in the cells and in the absence of sucrose, have given satisfactory results for the three types of lipases.

Differentiation of Lipases from Nonspecific Esterases

In order to differentiate between true lipases and nonspecific esterases it is necessary to use the inhibitory or promoting effect of different reagents.

A list of some inhibitors and accelerators of lipase and esterase activity is given in Table 1.

The addition to the incubation medium of selected promotors or inhibitors of lipase activity as well as of inhibitors of esterase activity will permit the cytochemist to differentiate between true lipases and nonspecific esterases.

Material fixed according to the described procedures and incubated in incubation media (3) to (6) allows us to visualize the presence of active lipases in plant cells.

Acid Lipase

This enzyme could be located in the membrane of spherosomes and in the glyoxysomal membrane of germinating spores (Figure 1). The activity of this enzyme could not be visualized after the first 2 days of germination.

Neutral Lipase

The incubation of specimens at pH 7.2 revealed the presence of this enzyme in the membrane of spherosomes, in the endoplasmic reticulum, and in dictyosomes of germinating spores. It was absent from glyoxysomes (Figures 5 and 6). Just as for acid lipase, this enzyme could not be visualized after 2 days of germination.

Alkaline Lipase

This enzyme could be located with great specificity in the glyoxysomal membrane, in mitochondria, and in thylakoids (Figure 2). Differentiation between active and inactive lipases was also possible (Figure 3). Two days after germination, this enzyme could no longer be detected in the glyoxysomes, but it persists in the thylakoids (Figure 4).

That the lead deposits correspond to the sites of activity of true lipases is shown by the control obtained by incubation in a medium containing quinine hydrochloride (inhibitor of lipase activity) and another containing sodium taurocholate (accelerator of lipase activity) (Figures 7 and 8).

128 *CRC Handbook of Plant Cytochemistry*

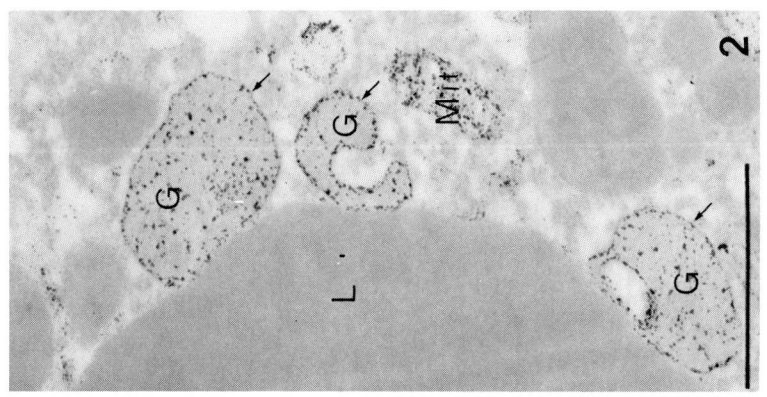

FIGURE 1. Acid lipase activity at 2 days of germination of *Bryum capillare* spores. Note the dense deposit around the lipids (L) (arrows) and at the microbodies (glyoxysomes) (G) (double arrows). Bar = 1 μm.

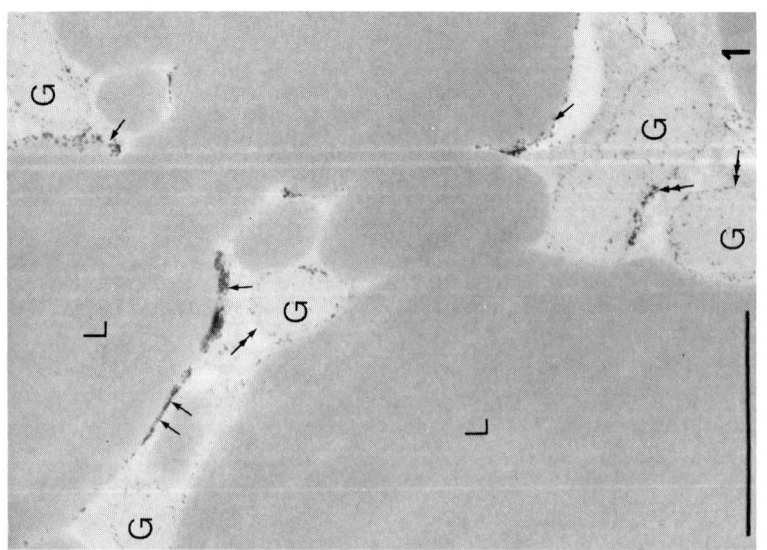

FIGURE 2. Alkaline lipase activity after incubation in a medium containing sodium taurocholate (accelerator of lipase activity) Note the presence of dense deposits at the membrane and the matrix of glyoxysomes (arrows). Reactivity is also evident at the mitochondria (Mit). Note the total absence of reactivity around the lipids. Bar = 1 μm.

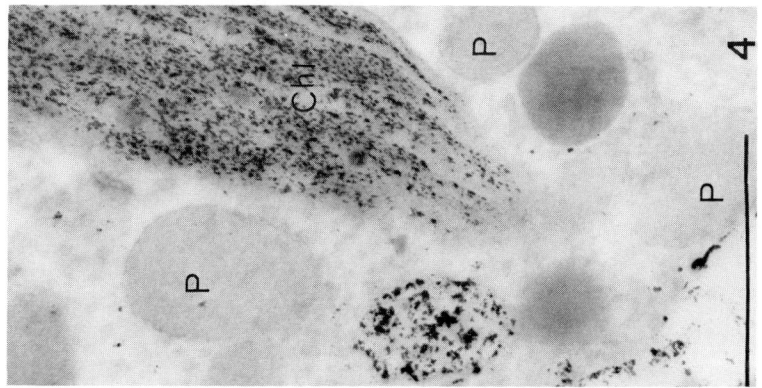

FIGURE 4. Detection of alkaline lipase in spores germinated during 4 days. No reactivity is seen at the microbodies (peroxisomes) or around the lipids. The reactivity of this enzyme in the thylakoids is evident. Bar = 1 µm.

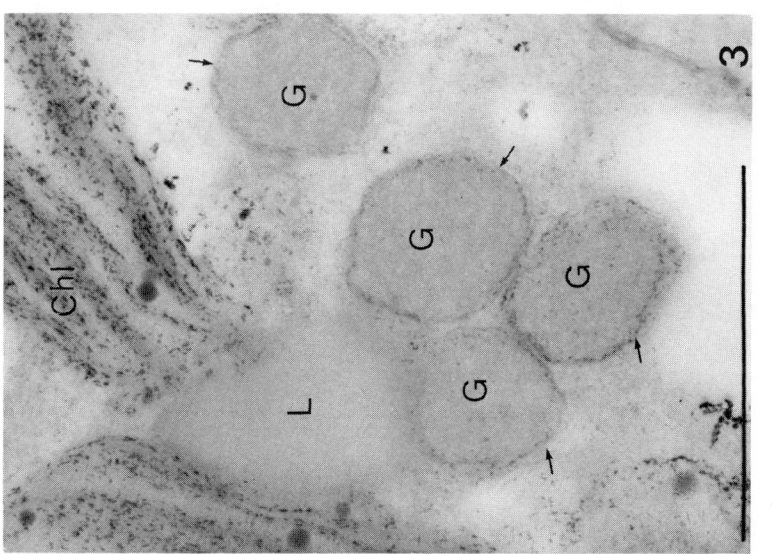

FIGURE 3. Alkaline lipase activity after incubation in a medium devoided of sodium taurocholate. Reactivity is visible in the glyoxysomal membrane and in thylakoids. This reactivity indicates the presence of active lipases at the glyoxysomal membrane and at the thylakoids. Comparing with Figure 2, the presence of inactive alkaline lipases at the glyoxysomal matrix and active lipases at the glyoxysomal membrane and at the thylakoids can be seen. Bar = 1 µm.

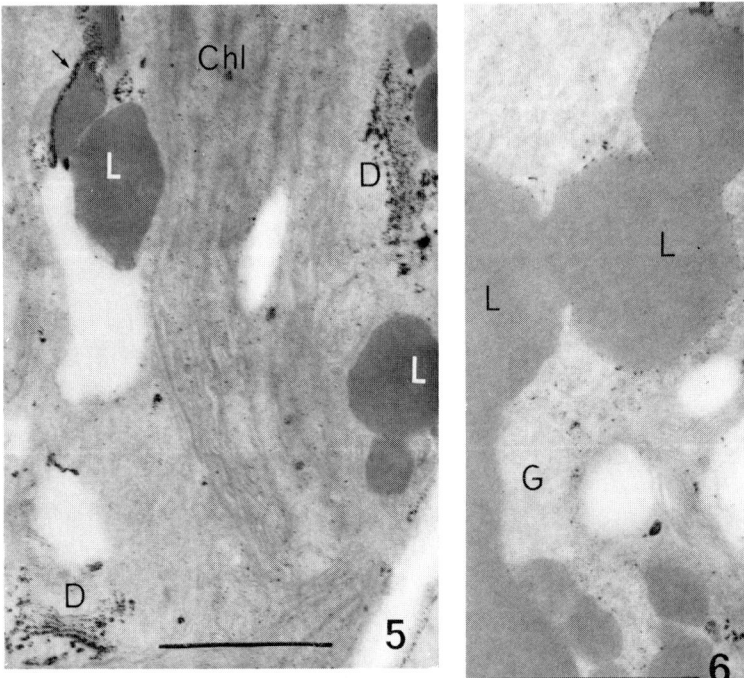

FIGURE 5 and 6. Neutral lipase activity at 2 days of germination. The reactivity of the endoplasmic reticulum (ER) and dictyosomes (D) is evident. No reactivity was detected at glyoxysomes (6) nor at the chloroplast thylakoids (5). Bar = 1 μm.

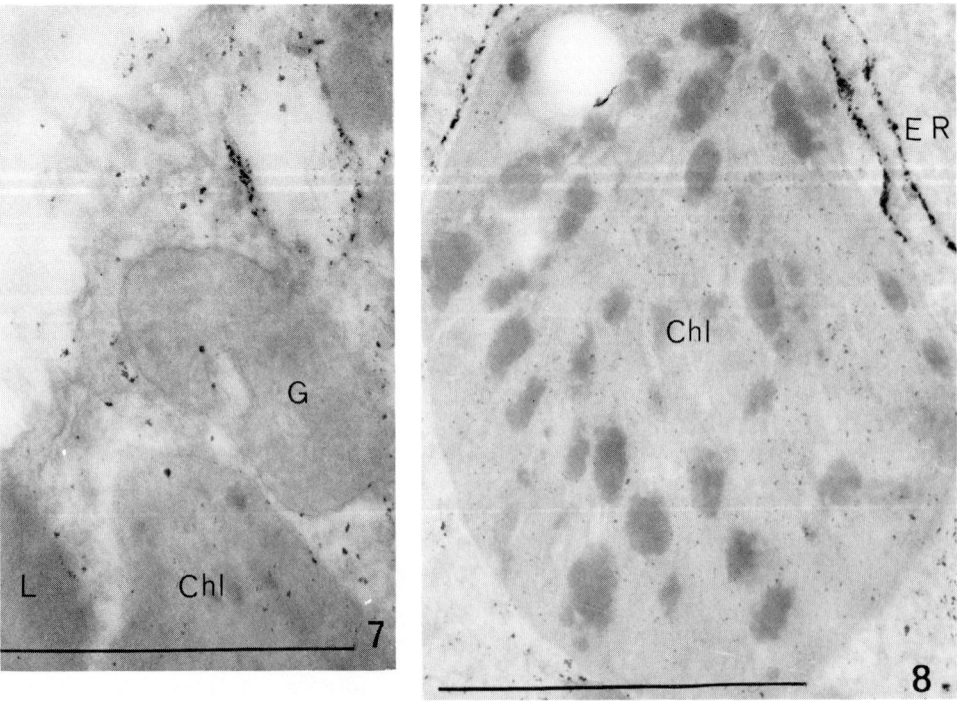

FIGURE 7 and 8. Controls obtained by inhibition of lipases by quinine hydrochloride. Note the total absence of reactivity around the lipid, at the glyoxysomal level (7), and at the thylakoids (8). These aspects demonstrate that the reactivity obtained at the organelles do correspond to the presence of true lipases at these levels. Bar = 1 μm.

REFERENCES

1. **Hanriot, M.,** Sur un nouveau ferment du sang, *C. R. Soc. Biol.,* 48, 925, 1896.
2. **Gomori, G.,** Histochemical localization of true lipases, *Proc. Soc. Exp. Biol. Med.,* 72, 697, 1949.
3. **Mark, D. D.,** Distribution of lipase in preneoplastic and neoplastic states induced in rat liver by paradimethylaminoazobenzene, *Arch. Pathol.,* 49, 545, 1950.
4. **Takeuchi, T., Furuta, M., and Yoshimira, K.,** Supplement of histochemical demonstration of tissue lipase, *Kumamoto Med. J.,* 9, 70, 1956.
5. **George, J. C. and Iype, P. T.,** Improved histochemical demonstration of lipase activity, *Stain Technol.,* 35, 151, 1960.
6. **Abe, M., Kramer, S. P., and Seligman, A. M.,** The histochemical demonstration of pancreatic-like lipase and comparison with the distribution of esterase, *J. Histochem. Cytochem.,* 12, 364, 1964.
7. **Takkar, G. L., Kambou, V. P., and Kar, A. B.,** Effect of altered hormonal states on the histochemical distribution of lipase activity in the rat prostatic complex, *Histochemie,* 20, 21, 1969.
8. **Mizuhira, V. and Kurotaki, A.,** Electron microscopic demonstration of lipase in the liver with Tween 40, *Niigata Med. J.,* 76, 651, 1962.
9. **Kramer, S. P., Aronson, L. D., Rosenfeld, M. G., Sulkin, M. D., Chang, A., and Seligman, A. M.,** Human pancreatic lipase study with bile salt activation and substrates from a homologous series of naphtyl alkanoates, *Arch. Biochem. Biophys.,* 102, 1, 1063.
10. **Hanker, J. S., Katzoff, L., Rosen, H. R., Seligman, M. L., and Seligman, A. M.,** Design and synthesis of thiolesters for histochemical demonstration of esterase and lipase via the formation of osmiophilic diazothioethers, *J. Med. Chem.,* 9, 288, 1966.
11. **Seligman, M. L., Veno, H., Hanker, J. S., Kramer, S. P., Wasserkrug, H., and Seligman, A. M.,** Cytochemical localization of pancreatic lipase with light and electron microscopy, *Exp. Mol. Pathol. Suppl.,* 3, 21, 1966.
12. **Murata, F., Yokota, S., and Nagata, T.,** Electron microscopic demonstration of lipase in the pancreatic acinar cells of mice, *Histochemie,* 13, 215, 1968.
13. **Nagata, T.,** Lipase, in *Electron Microscopy of Enzymes — Principles and Methods,* Vol. 2, Hayat, M. A., Ed., Van Nostrand Reinhold, New York, 1974, 132.
14. **Parshad, V. R. and Guraya, S. S.,** Morphological and histochemical observations on the digestive system of *Cotylophoron cotylophorum, J. Helmit.,* 52, 327, 1978.
15. **Takigami, S., Shikata, N., Murakami, Y., and Sotokichi, M.,** Retrials of ultracytochemical demonstrations of Tween lipase in pancrease, *J. Kansai Med. Univ.,* 2, 165, 1981.
16. **Schaffner, T., Elner, V. M., Bauer, M., and Wissler, R. W.,** Acid lipase: a histochemical and biochemical study using Triton X-100-naphtyl palmitate micelles, *J. Histochem. Cytochem.,* 26, 696, 1978.
17. **Bendayan, M., Roth, J., Perrelet, A., and Orci, L.,** Quantitative immunocytochemical localization of pancreatic secretory proteins in subcellular compartments of the rat acinar cell, *J. Histochem. Cytochem.,* 28, 149, 1980.
18. **Huang, A. H. C. and Moreau, R. A.,** Lipases in the storage tissues of peanut and other oil seeds during germination, *Planta,* 141, 111, 1978.
19. **Theimer, R. R. and Rosnitschek, I.,** Development and intracellular localization of lipase activity in rapeseed (*Brassica napus* L.) cotyledons, *Planta,* 139, 249, 1978.
20. **Donaldson, R. P.,** Organelle membranes from germinating castor bean endosperm. II. Enzymes, cytochromes and permeability of the glyoxysome membrane, *Plant Physiol.,* 67, 21, 1981.
21. **Singh, B., Sukhija, P. S., and Bhatia, I. S.,** Changes in total lipids, their components and lipases during seed germination, *Proc. Indian Nat. Sci. Acad.,* 47, 888, 1981.
22. **Lin, Y.-H., Moreau, R. A., and Huang, A. H. C.,** Involvement of glyoxysomal lipase in the hydrolysis of storage triacylglycerols in the cotyledons of soybean seedlings, *Plant Physiol.,* 70, 108, 1982.
23. **Furr, M. and Mahlberg, P. G.,** Histochemical analysis of laticifers and glandular trichomes in *Cannabis sativa, J. Nat. Prod.,* 44, 153, 1981.
24. **Chayen, J., Bitensky, L., Butcher, R., and Poulter, L.,** *A Guide to Practical Histochemistry,* Lippincott, Philadelphia, 1969, 46.
25. **Carrapico, F. and Pais, M. S.,** Cytochemical localization of lipases during spore germination from *Bryum capillare, Eur. J. Cell Biol.,* 30, 42, 1983.
26. **Liedvogel, B. and Kleining, H.,** Lipid metabolism in chromoplast membranes from daffodil: glycosylation and acylation, *Planta,* 133, 249, 1977.
27. **Gilkes, N. R., Herman, E. M., and Chrispeels, M. J.,** Rapid degradation of phospholipids in the cotyledons of mung bean seedlings, *Plant Physiol.,* 46, 38, 1979.
28. **Yoshida, S.,** Freezing injury and phospholipid degradation in vivo in woody plant cells. I. Sub-cellular localization of phospholipase D in living bark tissues of the black locust tree *(Robinia pseudacacia), Plant Physiol.,* 64, 241, 1979.

29. **Nakayama, Y. and Kito, M.,** Decomposition of soybean spherosomes by soybean phospholipase D, *Agric. Biol. Chem.,* 45, (9), 2155, 1981.
30. **Pugh, D. and Clawson, R. A.,** The cytochemical localization of phospholipase A and lysophospholipase in *Candida albicans, Sabouraudia,* 13, 110, 1975.
31. **Costa, A. L., Costa, C., Misefari, A., and Amato, A.,** On the enzymatic activity of certain fungi. VII. Phosphatidase activity on media containing sheep's blood of pathogenic strains of *Candida albicans, Atii Soc. Pelorit Sci. Fis. Nat.,* 14, 93, 1968.
32. **Blanchete-Mackie, Z. J. and Scow, R. O.,** Sites of lipoprotein lipase activity in adipose tissue perfused with chilomicrons, *J. Cell Biol.,* 51, 1, 1971.
33. **Molbert, E. R. G., Duspiva, F., and von Deimling, O. H.,** Die Histochemische Lokalisation der Phosphatase in der Tubulusepithelzelle der Mauseniere im Elektronmikroskopischen Bild, *Histochemie,* 2, 5, 1960.
34. **Adams, C. W. M., Abdulia, Y. H., Bayliss, O. B., and Weller, R. O.,** Histochemical detection of triglyceride esters with specific lipases and a calcium-lead sulphide technique, *J. Histochem. Cytochem.,* 14, 385, 1966.
35. **Buno, W. and Marino, R. G.,** Location of lipase activity in the chick embryo, *Acta Anat.,* 16, 85, 1952.
36. **Yoshimira, K.,** Histochemical studies on lipase, *Kumamoto Med. J.,* 29, 618, 1955.
37. **Diaconita, G.,** Untersuchungen uber die Anwendung von Tween substraten fur die Bestimmung der lipaseaktivitat, *Acta Histochem.,* 20, 82, 1965.
38. **Ogawa, K., Shinonaga, Y., and Suzuki, T.,** Metallophilia of the striated border of the rat jejunal epithelial cells, *Acta Anat. Nippon.,* 37, 134, 1962.
39. **Nachlas, M. M. and Seligman, A. M.,** Evidence for the specificity of esterase and lipase by the use of three chromogenic substrates, *J. Biol. Chem.,* 181, 343, 1949.
40. **Myers, D. K. and Mendel, B.,** Studies on ali-esterases and other lipid-hydrolising enzymes. I. Inhibition of the esterases and acetoacetate production of liver, *Biochem. J.,* 53, 16, 1953.
41. **Myers, D. K., Schotte, A., Boer, H., and Borsje-Bakker, H.,** Studies on ali-esterases and other lipid hydrolysing enzymes: inhibition of the esterases of pancreas, *Biochem. J.,* 61, 521, 1955.
42. **Spannhof, L. and Kreutzmann, H. L.,** Zur Verwendug von Naphtol AS carbonsaure Estern zum Histochemischen Nachweis von Esterasen und Lipasen, *Acta Histochem.,* 33, 394, 1969.

MALATE SYNTHASE

Richard N. Trelease

INTRODUCTION

Malate synthase (L-malate glyoxylate-lyase [CoA-acetylating], EC 4.1.3.2) is the second of two shunt enzymes in the anaplerotic glyoxylate cycle. It catalyzes the Mg^{2+}-dependent condensation of glyoxylate with the α-carbon of acetyl-CoA to form malate while liberating coenzyme A:

$$\underset{\text{Glyoxylate}}{{}^{-}\text{OOCCHO}} + \underset{\text{Coenzyme A}}{\text{CH}_3\text{COSCoA}} + \text{H}_2\text{O} \underset{}{\overset{Mg^{2+}}{\rightleftharpoons}} \underset{\text{malate}}{{}^{-}\text{OOCCH(OH) CH}_2\text{COO}^{-}} + \underset{\text{Coenzyme A}}{\text{COASH}} \quad (1)$$

Wong and Ajl[1] discovered the reaction in *Escherichia coli* cells growing on acetate as the sole carbon source. A year later Kornberg and Madsen[2] identified the significance of the reaction. Isocitrate lyase, which produces glyoxylate from the cleavage of isocitrate, works in conjunction with malate synthase to effect a by-pass or shunt of the tricarboxylic acid (TCA) cycle to replenish intermediates for growth on two-carbon compounds. This TCA cycle shunt is the glyoxylate cycle.

The glyoxylate cycle, with its two key component enzymes, was found soon thereafter to be an essential pathway in the conversion of storage lipids to carbohydrates in germinated castor beans.[3,4] It is now known that the glyoxylate cycle plays a pivotal role in the life cycle of numerous prokaryotes (bacteria) and eukaryotes including protozoans, nematodes, algae, fungi, mosses, ferns, and oil-rich seeds of higher plants (for reviews see references[5-10]). Unsubstantiated reports exist for the presence of malate synthase in certain invertebrates (e.g., see Skye and van Handel[11]), and until recently it was generally thought that the glyoxylate cycle was absent from vertebrates. Early claims for malate synthase activity in adult mammalian liver and kidney[12,13] were not verified, but recently activities of both malate synthase and isocitrate lyase were reported in fetal guinea pig liver[14] and in toad urinary bladder extracts.[15] In addition, malate synthase reactivity was demonstrated cytochemically (as described herein) in bladder epithelial cells.[15-17] These and other data indicate that the glyoxylate cycle is, indeed, operative in vertebrates, at least under certain conditions or at specific developmental stages.

In 1967, Breidenbach and Beevers[18] discovered in castor bean endosperm that the enzymes of the glyoxylate cycle were located exclusively in previously undefined organelles which they named "glyoxysomes". It is evident now that glyoxysomes are a specialized type of peroxisome, the latter being a generic term for organelles having a ubiquitous distribution among eukaryotes and being involved in a variety of metabolic pathways.[7-10] Malate synthase has been shown biochemically to be localized in glyoxysomes isolated from oil-storage tissues of numerous angiosperm seeds (Tables 2 and 3 from Huang et al.[10]), from several heterotrophic fungi and algae (Tables 4 to 9 from Huang et al.[10]), and from a few protozoans.[5] The only known exceptions for the occurrence of malate synthase in organelles other than peroxisomes comes from cell fractionation studies of two nematodes, *Ascaris suum* and *Turbatrix aceti*, where malate synthase and the other glyoxylate cycle enzymes were found in the mitochondria.[19-21] Cytochemical studies also indicate the localization of malate synthase in plant glyoxysomes.[10,22,23] Attempts were made to cytochemically elucidate the enzyme reactivity in isolated and *in situ Turbatrix* mitochondria, but genuine reaction product was not observed in the mitochondria or any other organelles.[41]

Several cytochemical procedures are available for detecting the chemical reactivity of enzymes within peroxisomes (e.g., catalase, oxidases, aminotransferases). However, these procedures define the localization of enzymes which are common to all types of peroxisomes, and, therefore, cannot be used to distinguish glyoxysomes from other peroxisomes having fundamentally the same morphology, involved in entirely different metabolic pathways. Isocitrate lyase, the other enzyme unique to glyoxysomes, has been localized in cotton glyoxysomes with an immunocytochemical procedure,[24] but not by a procedure exploiting its chemical reactivity. Thus, the procedure described below remains the only method by which one can ultrastructurally identify glyoxysomes *in situ* or in cell fractions without having a purified antigen and, hence, antibody to a specific glyoxysomal component. Isolation and purification of organelles from homogenized plant tissues is characteristically difficult, especially among the lower plants. Thus, malate synthase cytochemistry will continue to play an important role in elucidating the interactions of metabolic pathways within eukaryotic cells.

PARAMETERS OF THE ENZYME REACTION

The chemical basis for the cytochemical detection of malate synthase activity in tissue segments or organelle pellets is the reduction of ferricyanide by the sulfhydryl of coenzyme A in the presence of cupric ions to form the insoluble, electron-dense product cupric ferrocyanide, also known as Hatchett's brown. The overall reaction is shown below:

$$\text{acetyl-S-CoA} + \text{glyoxylate} \xrightarrow[\text{malate synthase}]{Mg^{2+}} \text{L-malate} + \text{CoA-SH} \quad (?)$$

$$2\ \text{CoA-SH} + 2\ \text{ferricyanide} \rightarrow 2\ \text{ferrocyanide} + \text{CoA-S-S-CoA} + 2H^+ \quad (3)$$

$$\text{ferrocyanide} + \text{cupric ion} \rightarrow \text{cupric ferrocyanide} \quad (4)$$

Trapping CoA-SH at the site of an enzyme for ultrastructural detection of enzyme activity was reported first by Higgins and Barnett[25] when localizing carnitine acetyltransferase in heart tissue. The concept of ferricyanide reduction by nonsulfhydryls in the presence of metal ions was employed first in light microscopic enzyme localizations,[26] then extended to the electron microscope. For background information and applications to plant enzymes, see Chapter 4 of this volume.

The successful application of the ferricyanide reduction method for localizing malate synthase required learning the behavior of the enzyme in the presence of the cytochemical reaction components. This was done, in part, by devising a spectrophotometric assay for malate synthase activity.[27] Ferricyanide has a broad absorption peak between 410 and 425 nm, whereas ferrocyanide absorbs strongly in the UV region as do the two substrates for the reaction, acetyl-CoA and glyoxylate.[22,27] Therefore, one can measure the substrate-dependent reduction of ferricyanide by CoA-SH at 420 nm without interference from other reaction components. The reduction of ferricyanide is linear with varying concentrations of sulfhydryl (such as cysteine) over a range of 5 to 50 nmol, yielding a micromolar extinction coefficient of 1.03 at 420 nm.[27]

Suitable concentrations of ferricyanide and the two substrates were determined with malate synthase activity in organelle pellets (11,000 $\times$ g, 30 min) centrifuged from homogenates of 3.5-day-old cucumber cotyledons.[28] Figure 1 shows the effect of varying ferricyanide concentrations on malate synthase activity. Concentrations less 75 nmol/mℓ limited the activity in the cuvette, but most importantly, enzyme activity was not affected by the presence

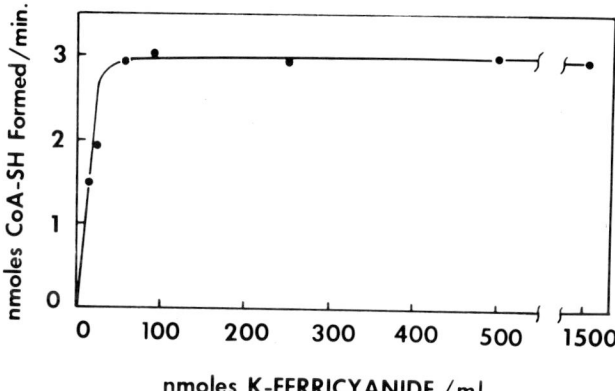

FIGURE 1. Malate synthase activity in the presence of varying potassium ferricyanide concentrations. The 1.0-mℓ reaction mixtures contained 60 μmol potassium phosphate (pH 7.6), 8 μmol MgCl$_2$, 0.2 μmol sodium glyoxylate, 0.25 μmol acetyl-CoA, and 13 μg protein (crude glyoxysomal pellet from cucumber cotyledons). (Reproduced from Trelease, R. N., Becker, W. M., and Burke, J. J., *J. Cell Biol.*, 60, 483, 1974. With permission.)

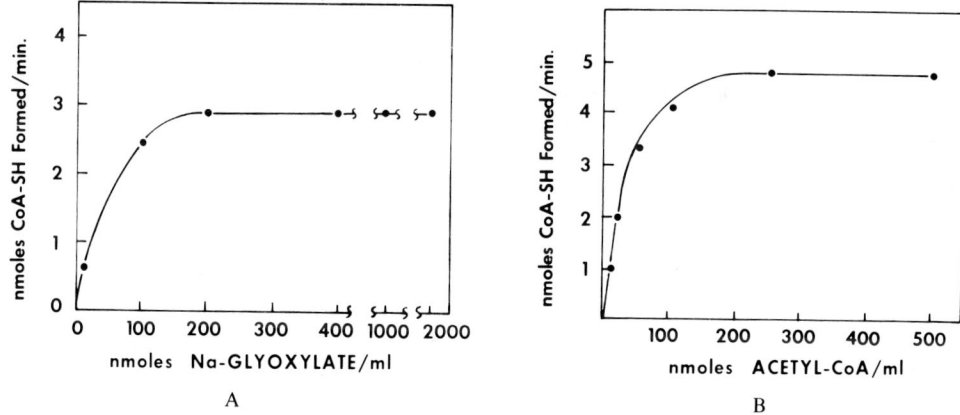

FIGURE 2. Malate synthase activity measured by ferricyanide reduction at 420 nm in presence of varying substrate concentrations. The 1.0-mℓ reaction mixtures contained 60 μmol potassium phosphate (pH 7.6), 8 μmol MgCl$_2$, and 0.1 μmol potassium ferricyanide. (A) Varying sodium glyoxylate concentrations: 0.25 μmol acetyl-CoA, 13 μg protein; (B) varying acetyl-CoA concentrations: 0.2 μmol sodium glyoxylate, 20 μg protein. (Reproduced from Trelease, R. N., Becker, W. M., and Burke, J. J., *J. Cell Biol.*, 60, 483, 1974. With permission.)

of potassium ferricyanide up to 1500 nmol/mℓ (1.5 mM). High concentrations of potassium ferricyanide give unacceptably high absorbances at 420 nm, therefore, assays for these experiments were conducted at 455 nm, using 0.15 as the micromolar extinction coefficient[29] to calculate activity. For other spectrophotometric assays, 100 nmol/mℓ (0.1 mM) ferricyanide was used and reduction by enzyme-released CoA-SH was measured at 420 nm. Figure 2A and B shows the effect of substrate concentration on malate synthase activity measured by ferricyanide reduction. Glyoxylate is inhibitory at concentrations above 4000 nmol/mℓ (4 mM) (data not shown). From these studies, final concentrations not limiting the enzyme activity, i.e., 0.1 mM potassium ferricyanide, 0.2 mM acetyl-CoA, and 0.5 mM sodium glyoxylate, were adopted for spectrophotometric assays.[27]

It is recommended, if feasible, for one to assay malate synthase activity from the test organism before performing the cytochemical procedure. Add the following components in the order given to a 1.1-mℓ cuvette (1-cm pathlength):

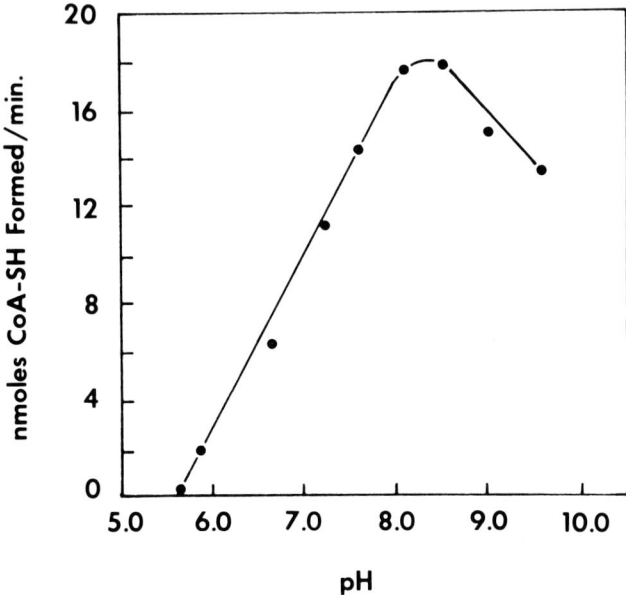

FIGURE 3. Activity of malate synthase in relation to pH. Assays were conducted with 50 μg protein of a glyoxysomal pellet prepared from 4-day-old etiolated cucumber cotyledons in 60 mM Tris-HCl and 60 mM potassium phosphate.

0.30 mℓ	200 mM Potassium phosphate (pH 7.6)
0.35 mℓ	Distilled H$_2$O
0.20 mℓ	40 mM MgCl$_2$
0.01 mℓ	10 mM Potassium ferricyanide
0.02 mℓ	10 mM Acetyl-CoA

Add extract (enzyme), then record any change in A$_{420}$ due to deacylase (hydrolysis of the acetyl-CoA which releases CoA-SH). Start the malate synthase reaction by adding 0.05 mℓ of 10 mM sodium glyoxylate. Vary the amount of water to accommodate variations in extract added to make a 1.0-mℓ reaction mixture. All reagents should be reagent grade. Acetyl-CoA can be purchased commercially, or due to its expense, one can simply acetylate commercial coenzyme A with acetic anhydride as follows. Dissolve 10 mg of coenzyme A in 1.0 mℓ of ice-cold 0.1 M aqueous sodium or potassium carbonate. Add 0.01 mℓ undiluted acetic anhydride, wait 5 min, and use. This is approximately a 10-mM solution. Store frozen between usage. Tris-HCl (pH 7.8) can be substituted for potassium phosphate in the spectrophotometric assay, but not the cytochemical reaction mixture (see later). Na, K-tartrate (up to 100 nmol per reaction mixture) did not have any effect on cucumber cotyledon malate synthase. Attempts to add copper sulfate solutions or copper chelated with tartrate in solution as a means of examining possible deleterious effects of copper on enzyme activity were unsuccessful in that a precipitate always formed with the crude enzyme extract in the cuvette. One may be able to test effects of copper in this assay with purified enzyme, if available. Similar problems of protein precipitation arise when one attempts to test possible consequences of fixatives (e.g., formaldehyde and/or glutaraldehyde) on enzyme activity.

Figure 3 shows the effect of pH on cucumber malate synthase activity. Its optimum activity in a cuvette is in the alkaline region at 8.3, although more than 50% of the peak activity is apparent at neutral pH. An in vitro optimum near pH 8.0 is the rule among the eukaryotic malate synthases recently examined.[30-33]

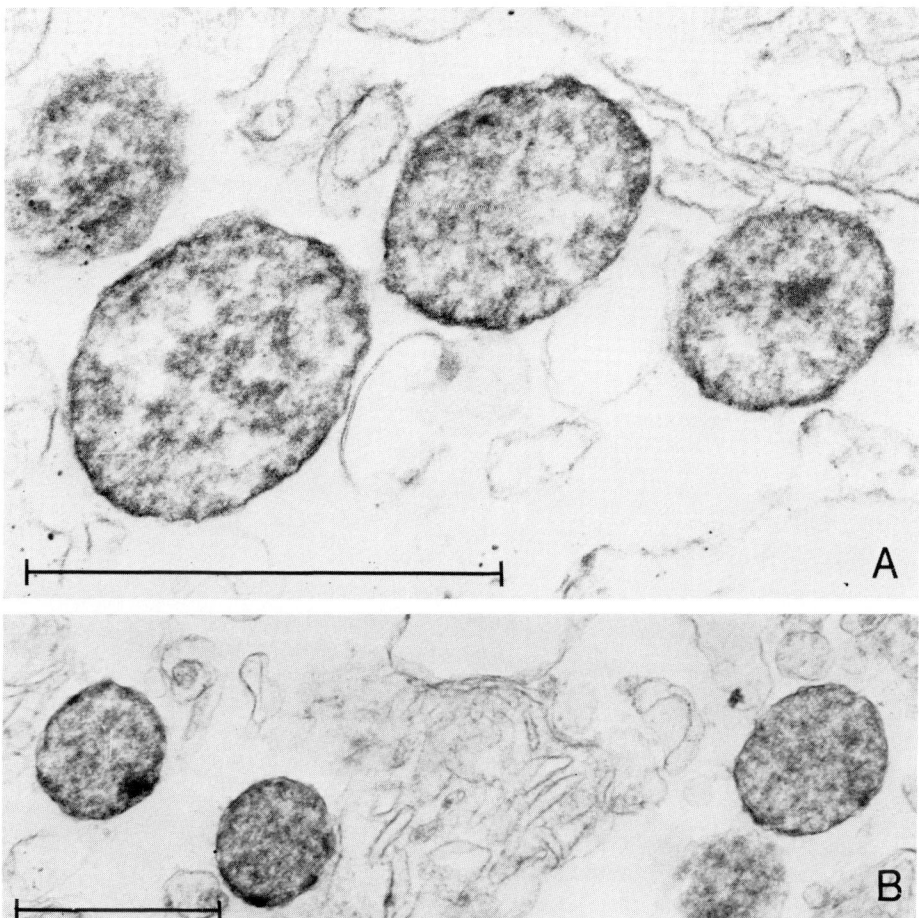

FIGURE 4. Yeast glyoxysomes recovered in 12,500 × g pellets[38] showing positive staining for malate synthase reactivity. Formaldehyde was used as a "prefixative" prior to incubation in the reaction mixture, then the organellar pellets were fixed in glutaraldehyde followed by osmium tetroxide. (A) Glyoxysomes from ethanol/methylamine-grown *Hansenula polymorpha*. Bar = 0.5 μm; (B) glyoxysomes from ethanol/methylamine-grown *Candida utilis*. Bar = 0.5 μm. Micrographs courtesy of M. Veenhuis.

CYTOCHEMICAL PROCEDURE

Fixation with Aldehydes

As with most cytochemical localizations of enzymes, fixation of the tissue or isolated organelles is a primary concern. With the exception of one study, glutaraldehyde alone or in combination with formaldehyde has been used as the fixative prior to incubation in the reaction mixture. Malate synthase in isolated yeast peroxisomes is completely inhibited by glutaraldehyde,[42] thus, a low concentration of only formaldehyde was used, the material incubated in the reaction mixture, then postfixed with glutaraldehyde to achieve suitable ultrastructural preservation (Figure 4A and B). A solution of 4% formaldehyde-1% glutaraldehyde or a 2 to 3% glutaraldehyde in either cacodylate or phosphate buffer (near pH 7.0) has been used successfully for a wide range of material.[15,16,23,27,34-38] Fixation times have varied from as little as 5 min (Figure 5) to as long as 4 to 5 hr (Figure 6); most fixations have been done at 4°C, although room temperature procedures seem equally effective.[22] Formaldehyde should be prepared fresh from paraformaldehyde, and purified glutaraldehyde (8 to 10%), rather than Practical grade (25%), is highly recommended. Attempts to incubate

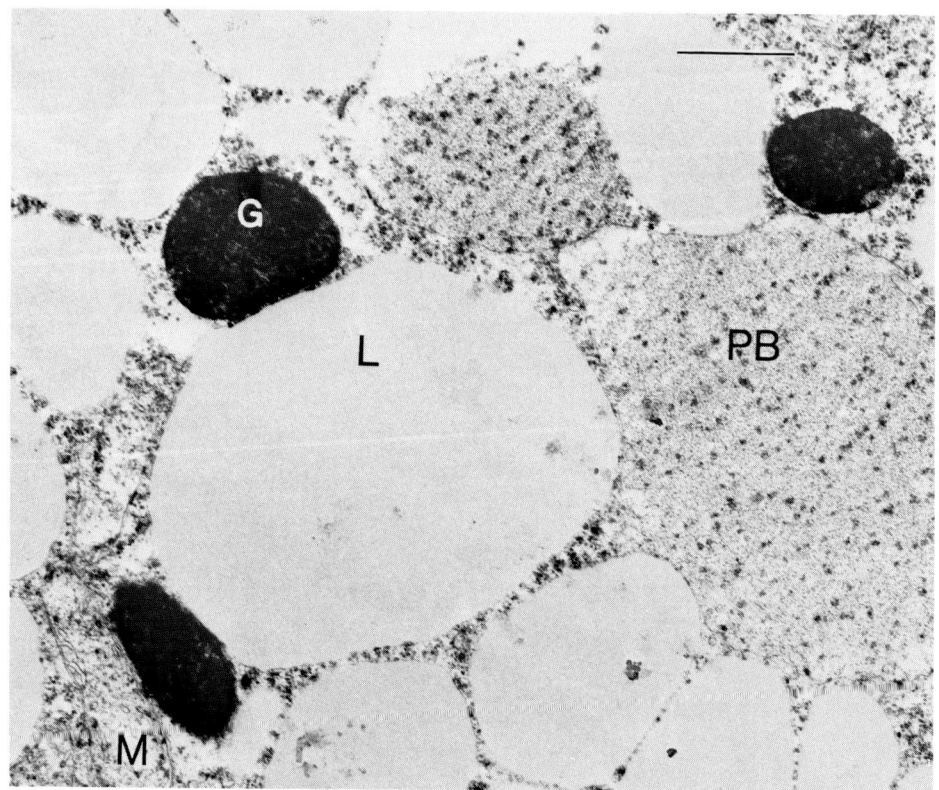

FIGURE 5. Portion of a cucumber cotyledon cell showing the appearance of glyoxysomes stained for malate synthase reactivity following a 5-min fixation (23°C) in a solution of 4% formaldehyde plus 1% glutaraldehyde in 50 mM sodium cacodylate, pH 7.1. Lipid bodies (L) and mitochondria (M) do not possess reaction product. Nonspecific background depositions are apparent in protein bodies (PB). Bar = 1.0 μm. (Magnification × 15,000.)

unfixed tissues in the reaction mixture (to eliminate possible aldehyde inhibitions) have not been successful, because some components of the reaction mixture apparently cannot pass through the unfixed plasma membrane (Figure 7).

Preincubation in Ferricyanide

Following fixation, the excess aldehyde should be washed out with several changes of fixation buffer at room temperature. The material then needs to be preincubated at least 30 min with gentle shaking or agitation in ferricyanide, preferably contained in low ionic-strength K-phosphate buffer at pH 6.9. I had previously recommended and used 3.0 mM potassium ferricyanide, which is twice the concentration in the final reaction mixture, to ensure oxidation of any components which could reduce the ferricyanide nonspecifically during incubation for malate synthase reactivity. Zwart et al.[38] experienced an overall staining of the cytoplasm in yeast spheroplasts, presumably due to an excess of reducing components in these cells. Vaughn[23] preincubated soybean cotyledon tissue in 20 mM potassium ferricyanide to oxidize endogenous reductants. This treatment did not inhibit the soybean enzyme and resulted in reduced background deposition (Figure 8). Thus, increasing the time and concentration of ferricyanide preincubation can be beneficial and should be tried if background depositions are a problem with individual tissues. Dimethyl sulfoxide (DMSO) (5%) may be used to enhance penetration of components during this step and during incubation in the reaction mixture. This has been done with cucumber cotyledons[27] and toad bladder epithelium.[16] Preincubating and incubating in the dark[23] may also reduce nonspecific ferricyanide reduction.

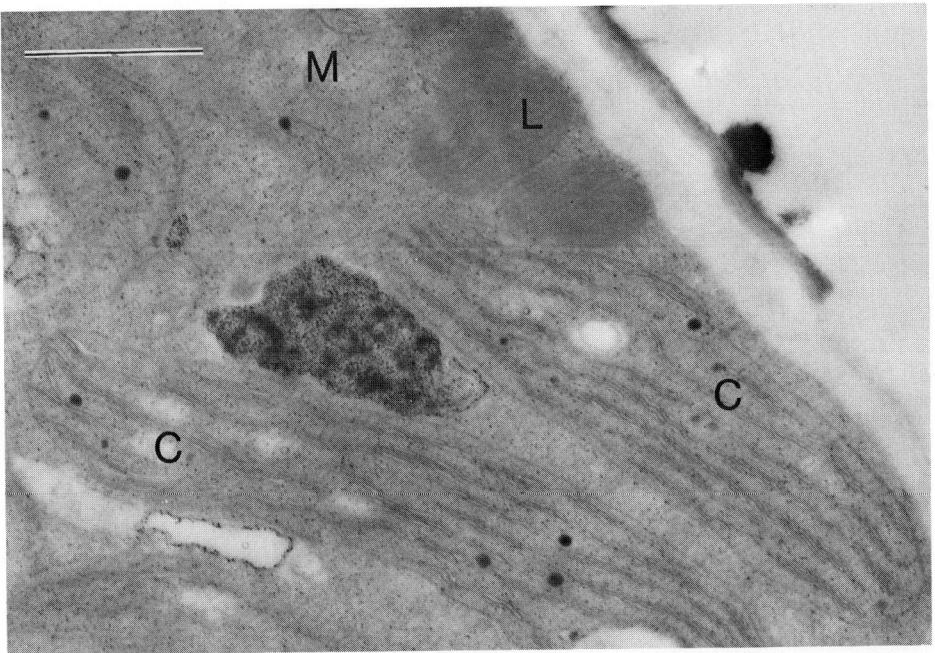

FIGURE 6. Portion of a cell in a *Bryum capillare* spore (4 days after sowing) fixed for 4.5 hr (4°C) in 2.5% glutaraldehyde in potassium phosphate (pH 6.8) illustrating reaction product attributable to malate synthase in a glyoxyperoxisome. The fixed spores were incubated in a reaction mixture containing 50 mM potassium phosphate at pH 8.5. A fine background deposition is observed in chloroplasts (C) and mitochondria (M), but not in lipid bodies (L). (Magnification × 24,000.) Bar = 1.0 μm. (From Pais, M. S. and Carrapico, F., *Ann. N.Y. Acad. Sci.*, 386, 510, 1982. With permission.)

Reaction Mixture

The tissue should be washed free of any ferricyanide reduced by cell components during the preincubation. This can be done by washing tissue in at least three changes of 10 to 20 mM phosphate buffer (pH 6.9), or fresh 1.5 mM potassium ferricyanide in the above buffer.

The following reaction mixture can be prepared during these final wash steps:

Volume (mℓ)	Stock solution	Final conc (mM)
0.30	65 mM K-phosphate (pH 6.9)	19.5
0.20	Copper-tartrate solution[a] (pH 6.9)	10 and 100
0.25	Distilled H$_2$O	—
0.03	50 mM Potassium ferricyanide	1.5
0.10	50 mM Magnesium chloride	5.0
0.02	150 mM Na-glyoxylate	3.0
0.10	10 mM Acetyl-CoA	1.0

[a] 50 mM copper sulfate and 500 mM Na,K-tartrate.

With the exception of the ferricyanide solution, all stock solutions can be prepared at least 1 day beforehand; the substrates are generally stored frozen. The copper-tartrate stock is made by slowly adding copper sulfate crystals to aqueous Na,K-tartrate with constant stirring and continued adjustment of pH to 6.9 with 1 N NaOH. This is done because unadjusted copper tartrate solution is near pH 5.0 and may decrease the final pH of the mixture to some

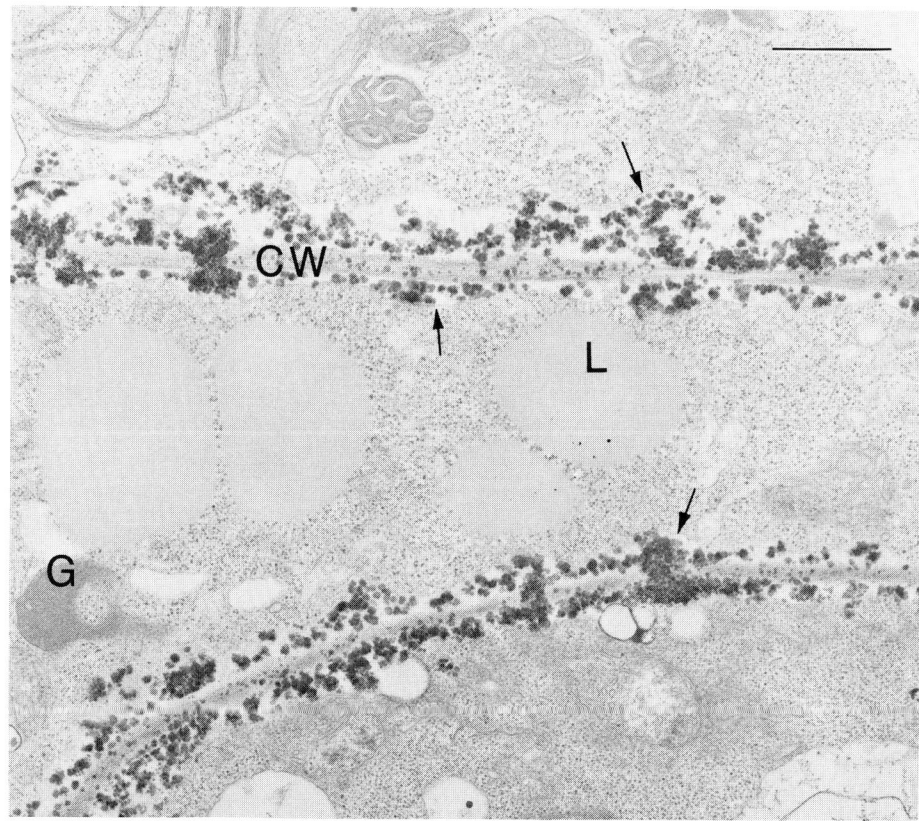

FIGURE 7. Portions of three cucumber cotyledon cells showing the results of incubating *unfixed* tissue segments in the malate synthase reaction mixture. Electron-dense material accumulates in areas between the cell walls (CW) and plasma membrane (arrows), but not in the cytoplasm or in organelles such as glyoxysomes (G). The tissue segments were fixed in formaldehyde-glutaraldehyde (4 to 1%) for 30 min following incubation, then postfixed in 2% osmium tetroxide. Bar = 2.0 μm. (Magnification × 8000.)

unknown value. For inclusion of DMSO, add 0.05 mℓ of stock DMSO after addition of the ferricyanide, and reduce the water to 0.20 mℓ. Final concentrations of the substrates and ferricyanide are higher than those used in the spectrophotometric assay discussed earlier. They were found not to be inhibitory in the cuvette reaction, hence, the higher concentrations are used for incubating tissue segments or pellets.

Tissue segments should be sliced from the parent tissue in approximately 1-mm squares during fixation for this incubation. Penetration of reactants into tissue segments is characteristically low, e.g., reaction product is seen only five to six cells (about 70 μm) in from the edge of oil seed cotyledon segments (Figure 9B). If incubating cell fractions or single-celled organisms, small (about 1 to 2 mm) pellets encased in 2% agar usually are suitable, but penetration of components likely will be limited to the outer edges of these pellets. The material should be placed in corked vials with a reduced air space, and incubated (with gentle shaking) at room temperature or 37°C for 30 to 90 min.

To my knowledge, no one has reported a significant variation in the original reaction mixture.[27] Jones et al.[16] decreased the final K-phosphate concentration from 19.5 to 6.0 mM to reduce background depositions and crystal formations (see later).

Following several washes (20 to 30 min) in 20 mM K-phosphate (pH 6.9), tissue segments or pellets are rinsed in buffer used for postfixation in 1 or 2% osmium tetroxide (1 to 2 hr). Dehydration may be done in a graded acetone or alcohol series (involving propylene oxide if desired) and embedded in Spurr's or an Epon resin. Sections may be double stained in

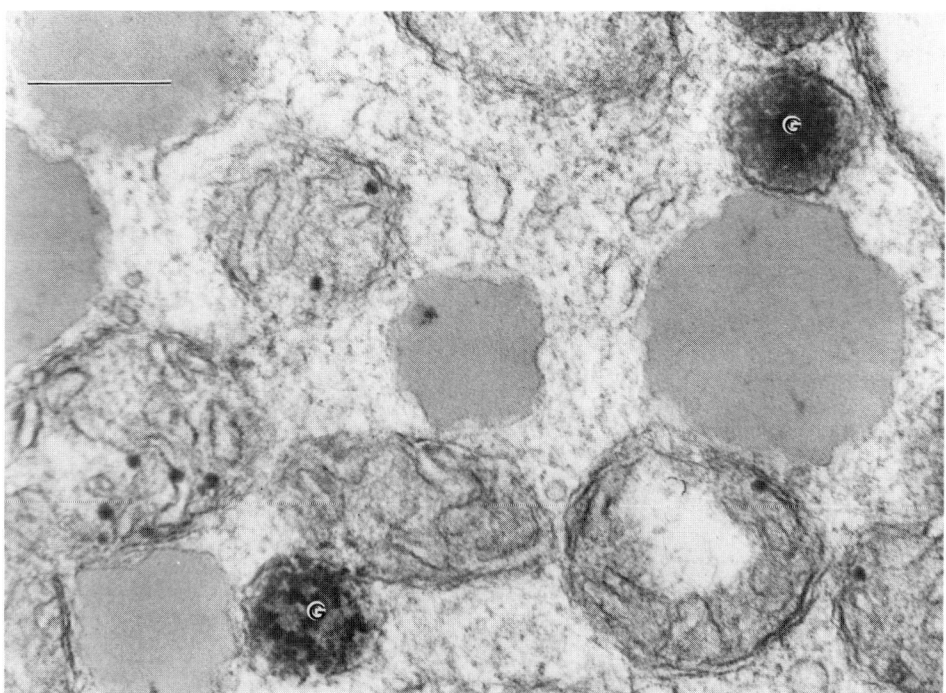

FIGURE 8. Malate synthase staining of glyoxysomes (G) in a soybean cotyledon cell. Note the absence of background deposition in organelles or the cytosol. Tissue segments were preincubated in 20 mM potassium ferricyanide prior to incubation in the reaction mixture. Bar = 0.5 μm. (From Vaughn, K. C., *Physiol. Plantarum*, 64, 1, 1985. With permission.)

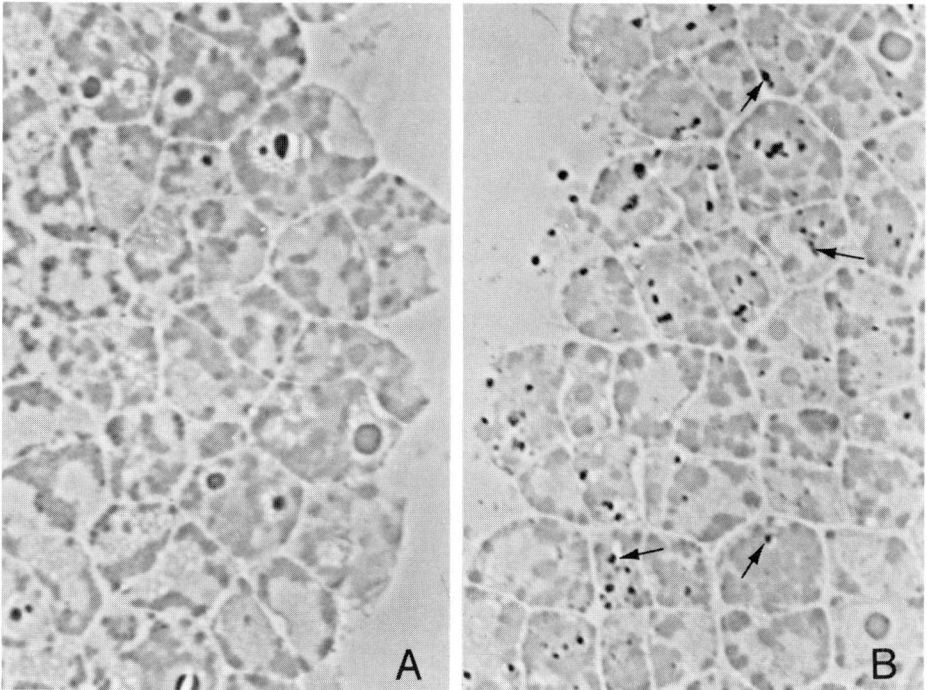

FIGURE 9. Phase micrographs of peridermally sectioned tissue segments of 3.5-day-old etiolated cucumber cotyledons incubated in the malate synthase reaction mixture. (A) Control segment, glyoxylate was omitted from the reaction mixture; (B) complete mixture, glyoxysomes are evident (arrows) as dark-stained structures in cells five to six cells from the edge of the tissue segment. (Magnification × 2000.)

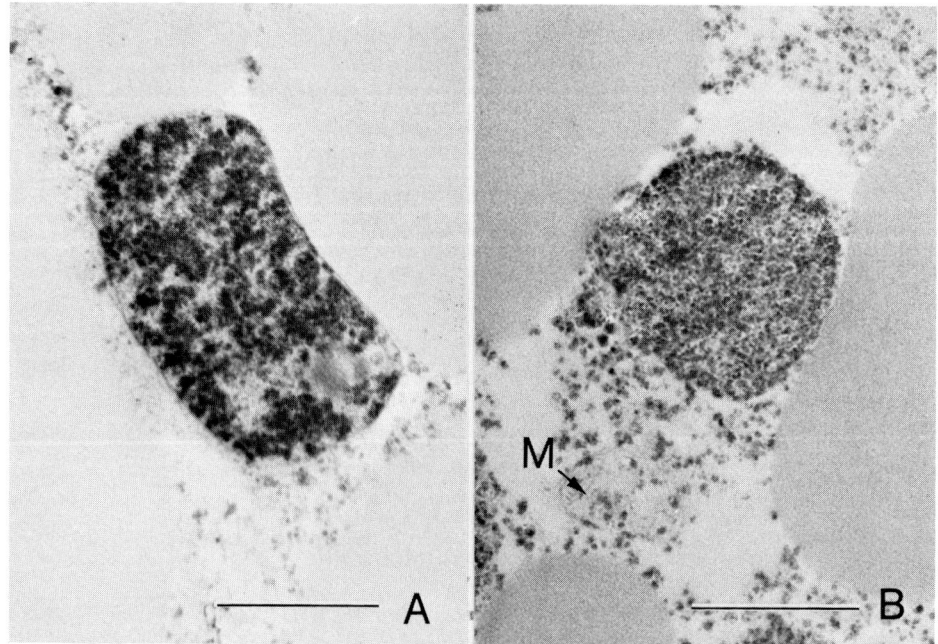

FIGURE 10. Reaction product attributable to malate synthase reactivity in oilseed glyoxysomes. (A) Glyoxysome in a sunflower cotyledon cell showing deposition of product nonuniformly in the matrix after poststaining in uranyl acetate and lead citrate. (Reproduced from Trelease, R. N., Becker, W. M., and Burke, J. J., *J. Cell Biol.*, 60, 483, 1974. With permission.) (B) Cucumber cotyledon glyoxysome with reaction product in the matrix without any poststain. M = mitochondria. Bars = 0.5 µm. (Magnification × 45,000.)

uranyl acetate and lead citrate (Figure 10A) or observed unstained (Figure 10B); no problems with poststaining have been reported.

The most critical control is to add acetyl-CoA and omit glyoxylate from the reaction mixture (Figures 9A and 11A). Cells may contain a nonspecific deacylase which would cleave acetyl-CoA giving a false interpretation of malate synthase reactivity.

Potential Problems and Modifications

The most common problems encountered with this cytochemistry are granular background depositions (Figures 11A and B and 12B) and nonenzymatic crystal formations (Figure 12A). The granular material likely is a copper precipitate and the crystal forms of magnesium phosphate. Magnesium is required for malate synthase activity,[33,39] and phosphate appears to be necessary for proper CoA-SH reduction of ferricyanide under these cytochemical conditions.[22,27,40] Crystal formation is especially evident at pH (7.6) and/or 60 mM phosphate. Reducing the pH and the magnesium and phosphate concentrations to those given in the reaction mixture generally eliminate the crystal formation. Lowering the pH to 6.9 may seem unreasonable in view of the in vitro pH optima being near pH 8.0 for malate synthases. However, if one examines the pH activity curve for cucumber malate synthase (Figure 3), it is apparent that at least 50% of the peak enzyme activity occurs at pH 7.0. Because cytochemical reactions are not quantitative, but depend only on maintaining activity under cytochemical conditions, the lower pH becomes an acceptable ploy to localize the enzyme. Preincubating tissue segments in high ferricyanide (20 mM) concentration, and thorough washes before and after incubation help reduce the granular depositions, but they are characteristic of copper ferrocyanide reduction methods (see Chapter 4 for other examples).

Several possible modifications of the method are discussed by Trelease,[22] but to my knowledge none of these have been attempted by others for localizing malate synthase reactivity.

MICROSCOPIC RESULTS

The reaction product in glyoxysomes of oilseed tissues is readily observed among lipid bodies with the optical microscope, both with brightfield illumination as dark brown material and, more explicitly, as dense material with phase microscopy (Figure 9B). The quality of deposition within the organelles or the degree of nonspecific background deposition usually cannot be determined with the optical microscope.

The amount and appearance of cupric ferrocyanide within the matrix of glyoxysomes vary considerably. The variability is not necessarily species dependent, because one may observe a range of deposition within the same tissue. For example, cupric ferrocyanide in the cucumber glyoxysomes shown in Figure 5 occupies nearly the entire matrix, exhibiting a nearly amorphous character, while the product shown in Figure 11B is more "patchy" and granular in appearance. The patchy or nonuniform appearance also is apparent in yeast (Figure 4A and B), fern spore (Figure 6), sunflower (Figure 10A), and *Entophlyctis* (Figure 13) glyoxysomes, whereas the reaction product is uniformly dense in the soybean glyoxysomes illustrated in Figure 8. No explanation has been given for these variations, i.e., they could depend on time for precipitation, distance from the edge of the tissue segment, penetrability of one or more critical reaction components into the cell or organelle, etc. It is this author's belief that one should be cautious in making physiological interpretations of the appearance of the product within the matrix. Cytochemistry of this nature is qualitative, demonstrating presence of enzyme activity (with proper controls) and subcellular localization of the enzyme.

One common feature of the cytochemical localization of malate synthase in the various tissues and species is the localization within the glyoxysomal matrix. This is of particular interest, because biochemical studies of malate synthase from several oilseed species show that the enzyme is a relatively hydrophobic molecule having an affinity for phospholipids, indicating that it is a membrane-associated enzyme.[8,10,30,32] Hence, the biochemical and cytochemical data seem to be conflicting. One possible explanation is that the enzyme is, indeed, membrane associated and the reaction product within the matrix is a consequence of product diffusing away from the enzyme site. This seems unlikely in that one would expect to see in some views of the numerous organelles examined a "ring" of product inside the membrane with the innermost interior of the matrix lacking product. No such views have been observed by this author or published by others. As it currently stands, no reconciliation of the cytochemical observations has been made with the biochemical information. Successful immunocytochemical localization of malate synthase in glyoxysomes from several species may contribute to the apparent disparity.

ACKNOWLEDGMENTS

This work was supported in part by NSF Grant DMB-8414857.

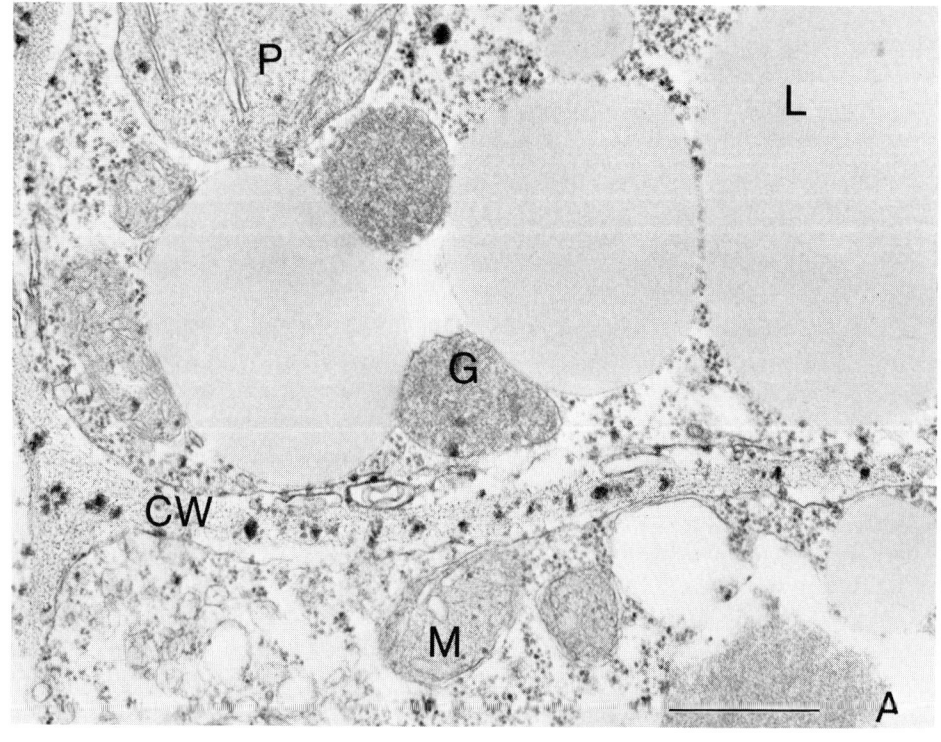

A

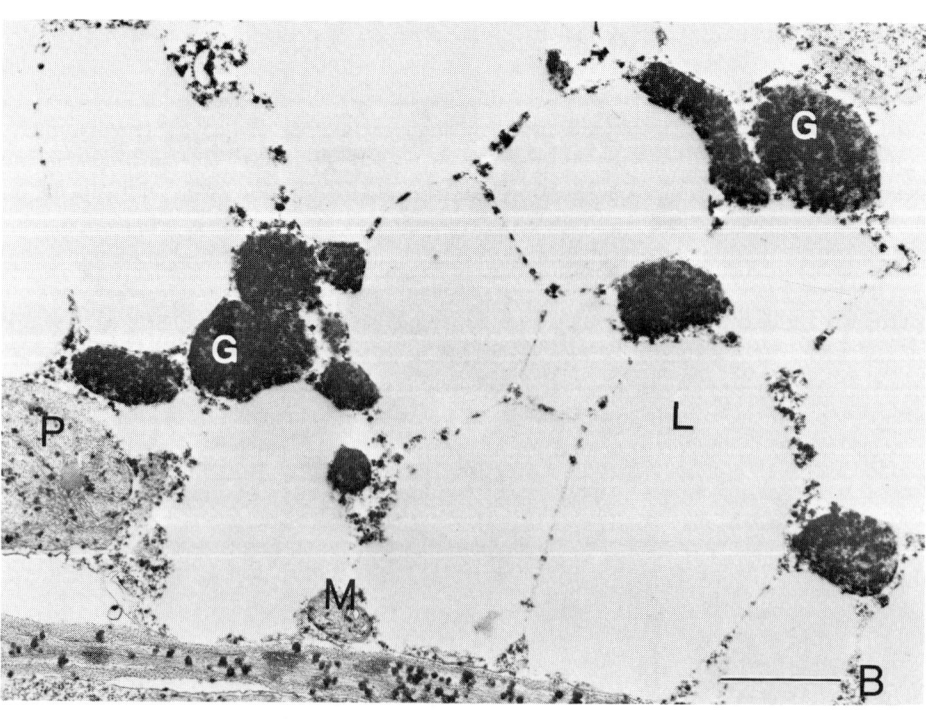

B

FIGURE 11. Portions of cucumber cotyledon cells. (A) Control segment, glyoxylate was omitted from the reaction mixture. Nonspecific granular deposits are in the cell wall (CW) and scattered in the cytoplasm. Magnification × 20,000; (B) complete mixture, granular material is concentrated in the glyoxysomal matrices (G) and lightly scattered in mitochondria (M) and plastids (P). L = lipid body. Bars = 1.0 μm. (Magnification × 16,500.)

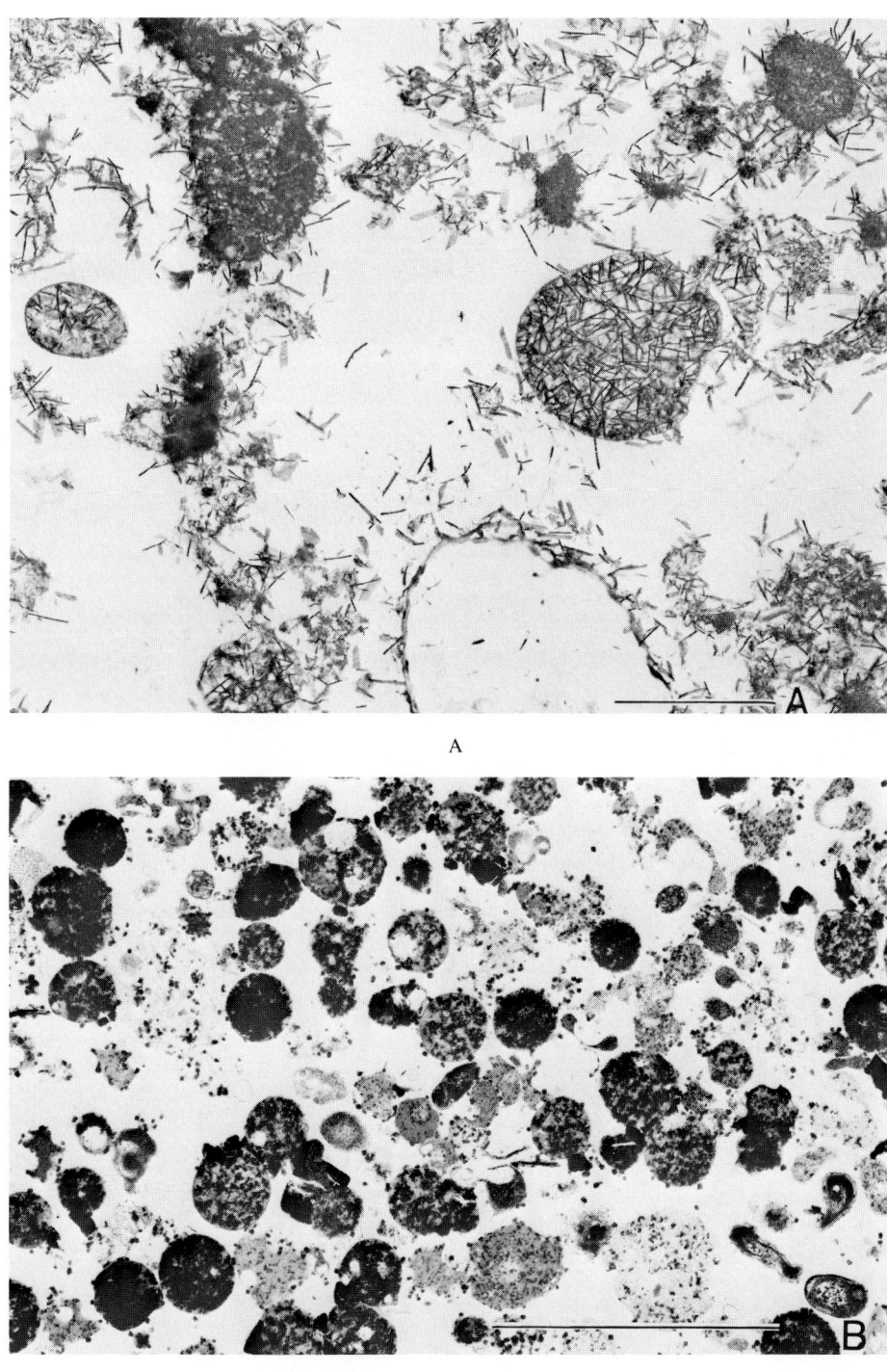

FIGURE 12. Oilseed glyoxysomes isolated in sucrose-density gradients,[34] then incubated for malate synthase reactivity. (A) Cottonseed glyoxysomes incubated in a reaction mixture containing 60 mM potassium phosphate, pH 7.1. Long crystal-like structures appear throughout the fraction and are concentrated in the glyoxysomes. Bar = 1.0 μm. (Magnification × 22,000.) (B) cucumber glyoxysomes incubated a mixture with 19 mM potassium phosphate, pH 6.9. Granular deposits occur in most organelles, but are concentrated in glyoxysomes. Bar = 5.0 μm. (Magnification × 8000.) (From Burke, J. J. and Trelease, R. N., *Plant Physiol.*, 56, 710, 1975. With permission.)

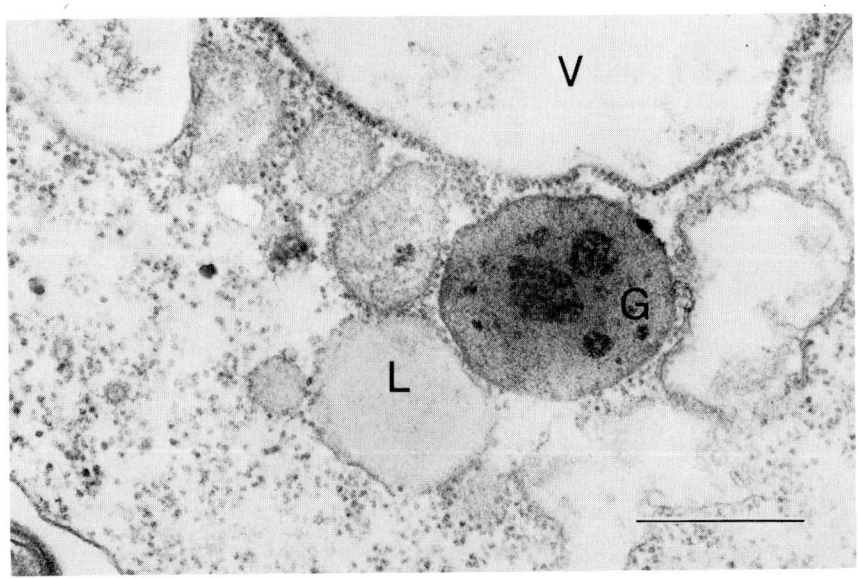

FIGURE 13. Appearance of reaction product in a glyoxysome of an encysted zoospore of *Entophlyctis variabilis*. Bar = 0.5 μm. (Magnification × 45,000.) (From Powell, M. J., *Protoplasma*, 89, 1, 1976. With permission.)

REFERENCES

1. **Wong, D. T. O., and Ajl, S. J.**, Conversion of acetate and glyoxylate to malate, *J. Am. Chem. Soc.*, 78, 3230, 1956.
2. **Kornberg, H. L. and Madsen, N. B.**, Synthesis of C_4-dicarboxylic acids from acetate by a "glyoxylate bypass" of the tricarboxylic acid cycle, *Biochem. Biophys. Acta*, 24, 651, 1957.
3. **Kornberg, H. L. and Beevers, H.**, A mechanism of conversion of fat to carbohydrate in castor beans, *Nature*, 180, 35, 1957.
4. **Yamamoto, Y. and Beevers, H.**, Malate synthetase in higher plants, *Plant Physiol.*, 35, 102, 1960.
5. **Müller, M.**, Biochemistry of protozoan microbodies: peroxisomes, α-glycerophosphate oxidase bodies, hydrogenosomes, *Ann. Rev. Microbiol.*, 29, 467, 1975.
6. **Gerhardt, B.**, *Microbodies/Peroxisomen pflanzlicher Zellen*, Cell Biology Monographs, Vol. 5, Springer-Verlag, Wien, 1978.
7. **Cioni, M., Pinzanti, G., and Vanni, P.**, Comparative biochemistry of the glyoxylate cycle, *Comp. Biochem. Physiol.*, B70, 1, 1981.
8. **Kindl, H. and Lazarow, P. B., Eds.**, Peroxisomes and glyoxysomes, *Ann. N.Y. Acad. Sci.*, 386, 1982.
9. **Huang, A. H. C.**, Metabolism in plant peroxisomes, *Recent Adv. Phytochem.*, 16, 85, 1982.
10. **Huang, A. H. C., Trelease, R. N., and Moore, T. S., Jr.**, *Plant Peroxisomes*, Academic Press, New York, 1983.
11. **Skye, G. E. and van Handel, E.**, Malate synthase in insects, *Comp. Biochem. Physiol.*, 49B, 83, 1974.
12. **Ganguli, N. C. and Chakraverty, K.**, Evidence for malic synthetase in animal tissues, *J. Am. Chem. Soc.*, 83, 2581, 1961.
13. **Liang, C.-C. and Ou, L.-C.**, Formation of malate from glyoxylate in animal tissues, *Biochem. J.*, 121, 447, 1971.
14. **Jones, C. T.**, Is there a glyoxylate cycle in the liver of the fetal guinea pig?, *Biochem. Biophys. Res. Comm.*, 95, 849, 1980.
15. **Goodman, D. B. P., Davis, W. L., and Jones, R. G.**, Glyoxylate cycle in toad urinary bladder: possible stimulation by aldosterone, *Proc. Natl. Acad. Sci. U.S.A.*, 77, 1521, 1980.
16. **Jones, R. G., Davis, W. L., and Goodman, D. B. P.**, Microperoxisomes in the epithelial cells of the amphibian urinary bladder: an electron microscopic demonstration of catalase and malate synthases, *J. Histochem. Cytochem.*, 29, 1150, 1981.

17. **Jones, R. G., Davis, W. L., and Goodman, D. B. P.,** The role of peroxisomes in the response of toad bladder to aldosterone, in *Peroxisomes and Glyoxysomes,* Vol. 386, Kindl, H. and Lazarow, P. B., Eds., 1982, 165.
18. **Breidenbach, R. W. and Beevers, H.,** Association of the glyoxylate cycle enzymes in novel subcellular particles from castor bean endosperm, *Biochem. Biophys. Res. Commun.,* 27, 462, 1967.
19. **Rubin, H. and Trelease, R. N.,** Subcellular localization of glyoxylate cycle enzymes in *Ascaris suum* larvae, *J. Cell Biol.,* 70, 374, 1976.
20. **McKinley, M. P. and Trelease, R. N.,** Glyoxylate cycle enzymes and catalase in digitonin-fractionated mitochondria in *Turbatrix aceti, Protoplasma,* 94, 249, 1978.
21. **McKinley, M. P. and Trelease, R. N.,** Regulation of carbon flow through the glyoxylate and TCA cycles in the mitochondria of *Tubatrix aceti*. I. Coarse and fine controls, *Comp. Biochem. Physiol.,* 67B, 17, 1980.
22. **Trelease, R. N.,** Malate synthase, in *Electron Microscopy of Enzymes, Principles and Methods,* Vol. 4, Hayat, M. A., Ed., Van Nostrand Reinhold, New York, 1975, 157.
23. **Vaughn, K. C.,** Structural and cytochemical characterization of three specialized peroxisome types in soybean, *Physiol. Plantarum,* 64, 1, 1985.
24. **Doman, D. C. and Trelease, R. N.,** Protein A-gold immunocytochemistry of isocitrate lyase in cotton seeds, *Protoplasma,* 124, 157, 1985.
25. **Higgins, J. A. and Barnett, R. J.,** Cytochemical localization of transferase activities: carnitine acetyltransferase, *J. Cell Sci.,* 6, 29, 1970.
26. **Karnovsky, M. J. and Roots, L.,** A "direct coloring" thiocholine method for cholinesterases, *J. Histochem. Cytochem.,* 12, 219, 1964.
27. **Trelease, R. N., Becker, W. M., and Burke, J. J.,** Cytochemical localization of malate synthase in glyoxysomes, *J. Cell Biol.,* 60, 483, 1974.
28. **Trelease, R. N., Becker, W. M., Gruber, P. J., and Newcomb, E. H.,** Microbodies (glyoxysomes and peroxisomes) in cucumber cotyledons. Correlative biochemical and ultrastructural study in light- and dark-grown seedlings, *Plant Physiol.,* 48, 461, 1971.
29. **Veeger, C., Der Vantarian, D. V., and Zeylemaker, W. P.,** Succinate dehydrogenase, in *Methods in Enzymology,* Vol. 13, Lowenstein, J. M., Ed., Academic Press, New York, 1969, 81.
30. **Koller, W. and Kindl, H.,** Glyoxylate cycle enzymes of the glyoxysomal membrane from cucumber cotyledons, *Arch. Biochem. Biophys.,* 181, 236, 1977.
31. **Woodcock, E. and Merrett, M. J.,** Purification and immunochemical characterization of malate synthase from *Euglena gracilis, Biochem. J.,* 173, 95, 1978.
32. **Riezman, H., Weir, E. M., Leaver, C. J., Titus, D. E., and Becker, W. M.,** Regulation of glyoxysomal enzymes during germination of cucumber. III. *In vitro* translation and characterization of four glyoxysomal enzymes, *Plant Physiol.,* 65, 40, 1980.
33. **Miernyk, J. A. and Trelease, R. N.,** Malate synthase from *Gossypium hirsutum, Phytochemistry,* 20, 2657, 1981.
34. **Burke, J. J. and Trelease, R. N.,** Cytochemical demonstration of malate synthase and glycolate oxidase in microbodies of cucumber cotyledons, *Plant Physiol.,* 56, 710, 1975.
35. **Powell, M. J.,** Ultrastructure and isolation of glyoxysomes (microbodies) in zoospores of the fungus *Entophylyctis* sp., *Protoplasma,* 89, 1, 1976.
36. **Pais, M. S. and Carrapico, F.,** Localization cytochimique de la malate synthétase et de la glycolate oxydase aū niveau des microbodies des spores chlorophylliennes de la mousse *Bryum capillare, C.R. Acad. Sci. Paris,* 288, 395, 1979.
37. **Pais, M. S. and Carrapico, F.,** Microbodies — a membrane compartment, *Ann. N.Y. Acad. Sci.,* 386, 510, 1982.
38. **Zwart, K. B., Veenhuis, M., Plat, G., and Harder, W.,** Characterization of glyoxysomes in yeasts and their transformation into peroxisomes in response to changes in environmental conditions, *Arch. Microbiol.,* 136, 28, 1983.
39. **Eggerer, H. and Klette, A.,** Über das Katalyseprinzip der Malate-synthase, *Eur. J. Biochem.,* 1, 447, 1967.
40. **Lukaszyk, A.,** A method for histochemical demonstration of α-glycerophosphate-ferricyanide oxidoreductase activity, *Folia Histochem. Cytochem.,* 9, 167, 1971.
41. **McKinley, M.,** personal communication.
42. **Veenhuis, M.,** personal communication.

PECTINASE

Craig L. Nessler and Randy D. Allen

INTRODUCTION

Pectins comprise a complex class of polysaccharides which are particularly abundant in the middle lamella of plant cell walls where they function in the adhesion of adjacent cells. The major component of the pectins are straight chain molecules of 1,4-α-D-polygalacturonic acid. Many of the galacturonan subunits in pectins are methyl esterified, however, the remaining residues are negatively charged and can cross-link the polymers into a semirigid gel by binding Ca^{2+}.

Pectinase (poly-[1,4-α-D-galacturonide glycanohydrolase]; EC 3.2.1.15) is an enzyme that randomly hydrolyzes the α 1 → 4 linkages of pectin. This enzyme has been implicated in several important developmental processes in vascular plants. Pectinases are associated with the softening of fleshy fruits during the ripening process.[1] Abscission of leaves[2,3] and reproductive structures[4,5] may also involve pectin hydrolysis to allow for the separation of cells in abscission layers. The penetration of pollen tubes through stylar tissue[6] and the intrusive growth of nonarticulated laticifers[7,8] both appear to be facilitated by the enzymatic removal of middle lamella between cells ahead of the growing tip.

A method for the ultrastructural localization of pectinase activity in plant tissues has recently been developed.[8] This procedure is a modification of a technique described by Bal[9] for the *in situ* identification of cellulases. Using our procedure we have successfully localized pectinase in the growing tips of laticifers in *Nerium oleander*. The pectinase localization technique described here should be of use in the study of other systems of intrusive growth and may also be applicable to developmental processes involving schizogeny.

RATIONALE

Benedict's solution is a commonly used reagent for the qualitative detection of reducing sugars. When heated in the presence of reducing sugars, Benedict's solution changes from clear blue to cloudy red with the formation of an insoluble precipitate. This precipitate is cuprous oxide (Cu_2O) crystals which are formed by the reduction of cuprous salts by sugar aldehydes.

The pectinase localization procedure described here takes advantage of the electron opacity of Cu_2O crystals. Benedict's solution can, therefore, be used to localize galacturonic acid residues which are liberated from exogenously supplied pectin during enzymatic hydrolysis (Figure 1).

Electron opaque reaction product in a laticifer vacuole was examined by energy dispersive X-ray analysis (Figure 2). Copper was identified in a crystalline deposit (Figure 3, spectrum A), but was absent from an adjacent area of the same specimen (Figure 3, spectrum B) as predicted (Figure 1). It is interesting to note the presence of osmium in the inclusion that is deposited during postfixation and serves to increase electron density of the reaction product.

PROCEDURE

Fixation

Specimens are fixed for 1 to 2 hr on ice by immersion in a modified Karnovsky fixative[10] composed of 3% paraformaldehyde and 2% glutaraldehyde in 0.05 M phosphate buffer, pH 7.2. The osmolality of the fixative solution may be adjusted to prevent plasmolysis in a particular tissue. Fixation is performed on ice using blocks of tissue 0.5 to 1.0 mm^3.

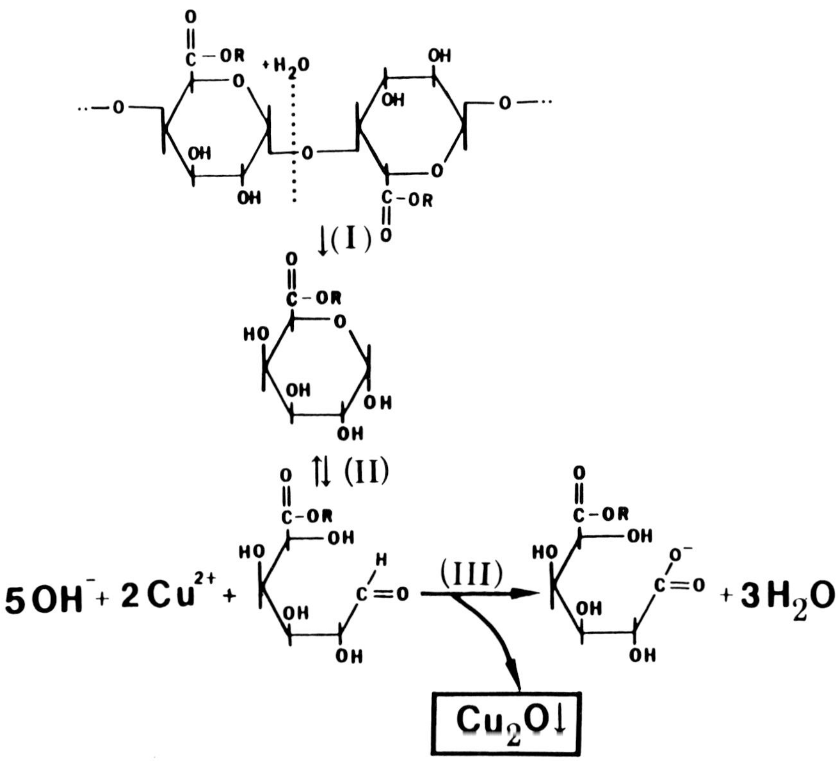

FIGURE 1. Theoretical basis for the pectinase localization procedure. R = H (galacturonic acid); R = CH_3 (methyl esterified galacturonan). (I) Enzymatic hydrolysis of pectin by pectinase; (II) equilibrium between hemiacetyl and noncyclic aldehyde forms of reducing sugar; (III) formation of cuprous oxide precipitate by reduction of cuprous ion by reducing sugar.

Washing

Specimens must be thoroughly washed with at least 20 changes of ice-cold 0.05 M phosphate buffer (pH 7.2) to prevent the reaction of Benedict's solution with aldehyde groups in the fixative. After washing, tissues are further equilibrated with the same buffer by storing them overnight at 0°C.

Substrate Incubation

Tissues are placed into a substrate solution composed of 0.5% pectin (Sigma Chemical Co. No. p-9135) in 0.1 M sodium acetate buffer, pH 5.0. Specimens are incubated in this solution at room temperature for 20 min, during which time pectin molecules diffuse into the tissue and are hydrolyzed *in situ* to release reducing sugars. The time of incubation can be varied, but should remain relatively short to minimize the diffusion of galacturonic acid residues away from sites of enzyme activity.

Reaction Product Formation

Following incubation in substrate, specimens are immediately transferred to Benedict's solution and boiled for 10 min. This stops the enzymatic reaction and results in the formation of electron-opaque cuprous oxide reaction product. The tissues are allowed to cool, rinsed with several changes of 0.05 M sodium cacodylate buffer (pH 7.2), and then postfixed for 2 hr in 1% osmium tetroxide in the same buffer. Specimens are rinsed with water and dehydrated through a graded ethanol series into acetone or propylene oxide. Both Spurr's[11] low-viscosity resin or Epon-Araldite have been successfully used, although the latter seems to provide slightly better contrast.

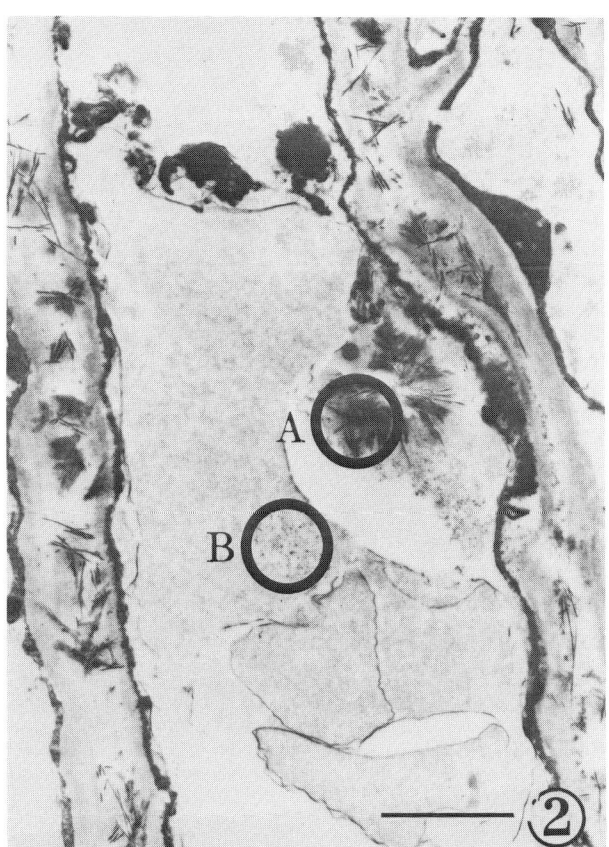

FIGURE 2. Section through experimentally treated tissue of oleander that was used for X-ray analysis. Electron opaque pectinase reaction product (A) and adjacent deposit-free area (B) in laticifer vacuole. Bar = 2 μm.

Controls

In order to interpret the pattern of reaction product deposition in experimental tissues, a series of controls must be processed and examined. These controls should include the following.

Minus substrate — The specificity of localization procedure depends on the ability of tissue pectinases to liberate galacturonic acid residues from exogenously supplied pectin. Because Benedict's solution will react with endogenous reducing sugars, as well as the reducing ends of other polysaccharides, it is important to determine the background level of reaction product formed in the absence of added substrate. For this control, tissues are fixed and washed as outlined above, but are incubated in acetate buffer without pectin prior to boiling in Benedict's solution.

Heat inactivation — An enzymatic basis for pectin degradation can be assumed if reaction product is absent or greatly diminished in control specimens in which enzymes have been heat denatured. Heat inactivation can be accomplished for this control by boiling tissues for at least 10 min just prior to substrate incubation.

Minus Benedict's solution — Although the appearance of the Cu_2O crystals formed during the Benedict's reaction are quite distinct, it is recommended that additional controls be performed in which Benedict's solution is omitted. Specimens should, therefore, be examined which have been soaked in pectin, but boiled in buffer instead of Benedict's solution.

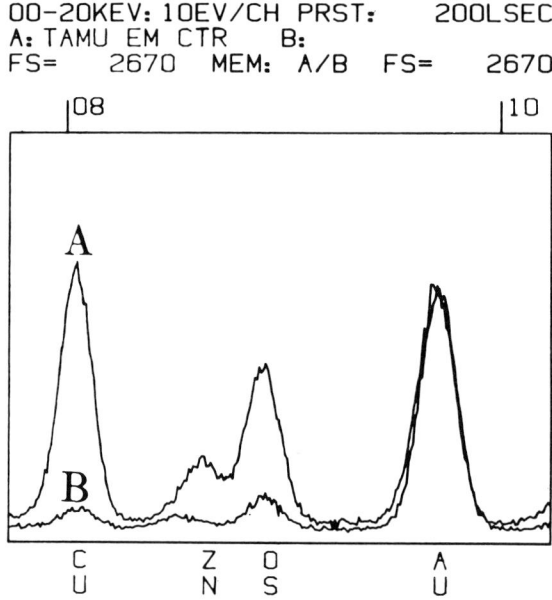

FIGURE 3. Energy dispersive X-ray microanalytical spectra of section in Figure 2 showing high levels of copper and osmium in electron opaque reaction product (A) compared with adjacent deposit-free area (B). Full scale = 2670 counts for both spectra.

Staining

Thin sections of all treatments can be viewed unstained, however, traditional heavy metal stains provide additional contrast without obscuring reaction product. Excellent results have been obtained by staining sections with aqueous uranyl acetate for 10 min at 40°C followed by 10 min in Reynold's[12] lead citrate at room temperature.

APPLICATIONS

Nonarticulated laticifers have been described as growing intrusively at their tip followed by symplastic elongation of the rest of the cell.[13] Wilson et al.[7] identified pectinase activity in freshly isolated latex of *Asclepias syriaca* and suggested that pectinase secreted by the growing laticifer could loosen the middle lamella between adjacent parenchyma cells to permit passage of the laticifer tip between cells. Pectinase might also serve to loosen the middle lamella between laticifers and surrounding cells to facilitate laticifer elongation. The procedure described here has been used to localize pectinase activity in the nonarticulated laticifers of *Nerium oleander*.[8]

In the growing tips of oleander laticifers reaction product appears as electron-opaque, crystalline deposits along the middle lamella and in small, dense patches within the cytoplasm (Figure 4). Mature laticifers contain abundant deposits in the central vacuole in addition to those along the middle lamella (Figure 5). Reaction product is generally concentrated into localized masses associated with patches of cytoplasm (Figures 6 and 7), but occasionally appears more dispersed (Figure 8), probably due to diffusion of galacturonan residues prior to their reaction with Benedict's solution.

Dense accumulations of reaction product are absent from the laticifer vacuoles of tissues incubated in buffer alone (Figure 9) and in boiled controls (Figure 10). In both controls, however, the electron-opaque deposits seen in the middle lamella of laticifer walls are still recognizable. The presence of reaction product in the middle lamella of laticifers in controls may be due to hydrolysis of the endogenous pectin in this region during incubation in buffer prior to boiling.

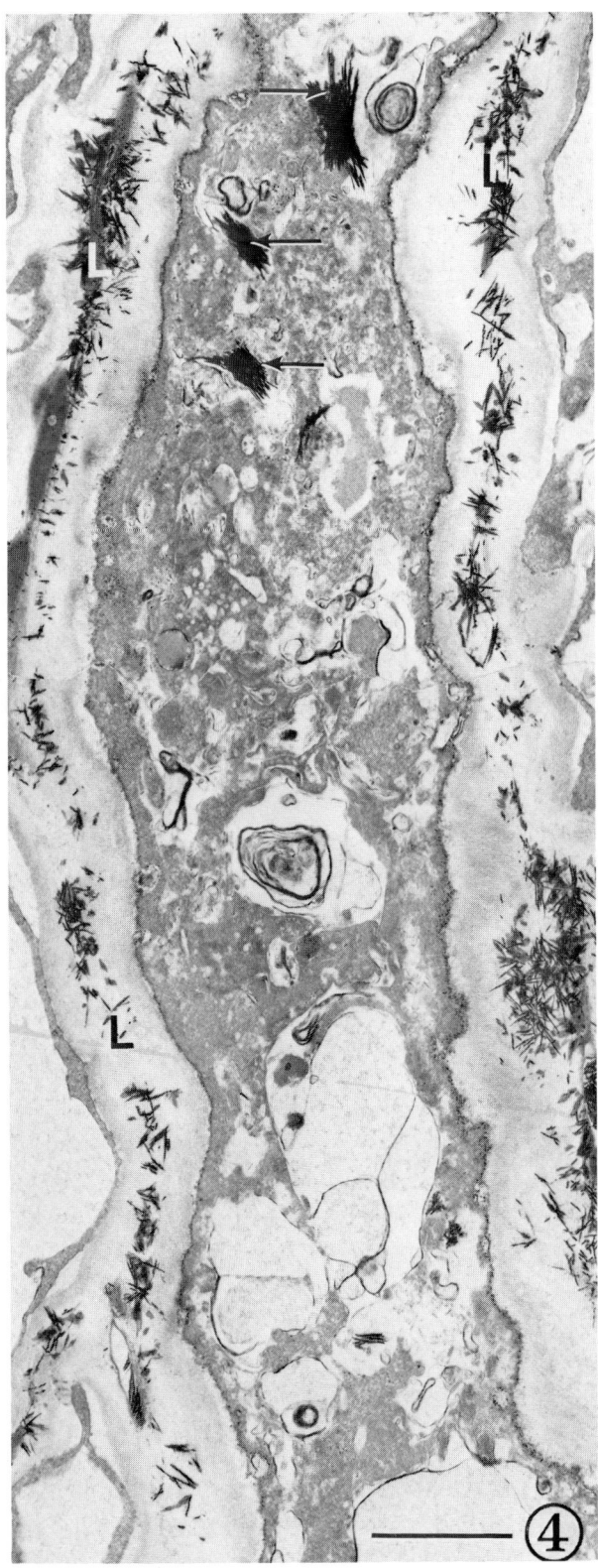

FIGURE 4. Immature oleander laticifer with extensive pectinase reaction product in middle lamella (L) and smaller cytoplasmic deposits (arrows). Bar = 2 μm.

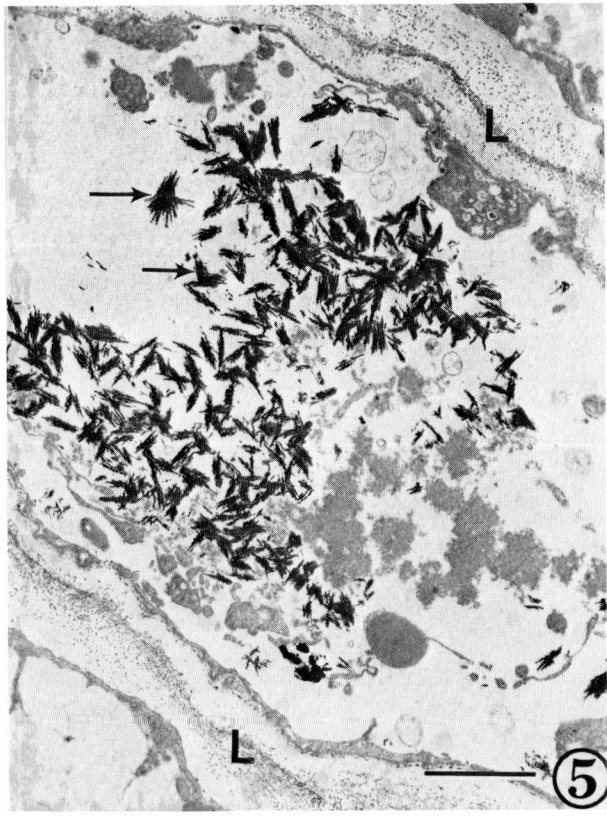

FIGURE 5. Electron opaque reaction product (arrows) showing pectinase activity in vacuole of laticifer and middle lamella (L). Bar = 2 μm.

ACKNOWLEDGMENTS

The authors wish to thank Dr. Robert Burghardt, Director of the Texas A & M University Electron Microscopy Center, for his assistance in performing the energy dispersive X-ray analysis. Funding for this research was provided by NSF Grant PCM-8025003 to C.L.N.

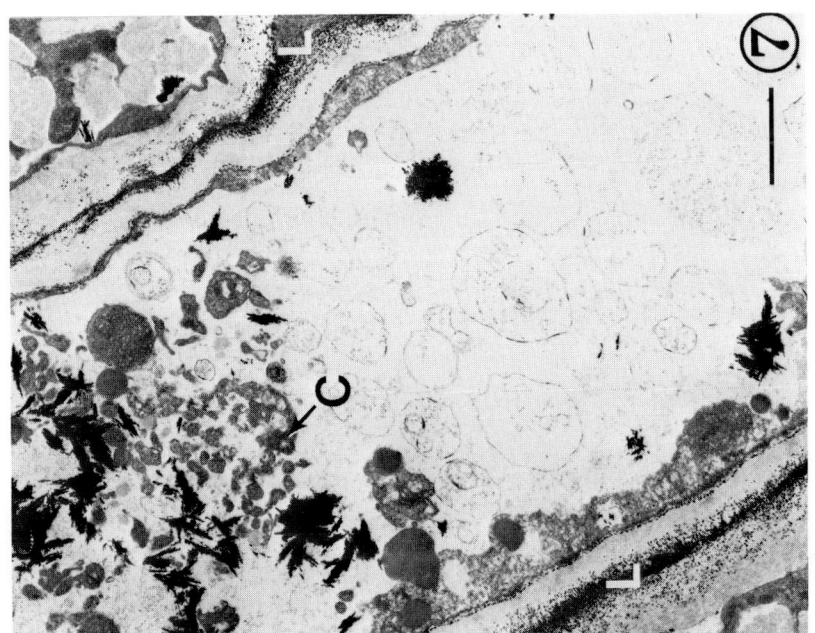

FIGURE 7. Reaction product localizing pectinase activity in areas of laticifer cytoplasm (C) within the central vacuole. Bar = 2 μm.

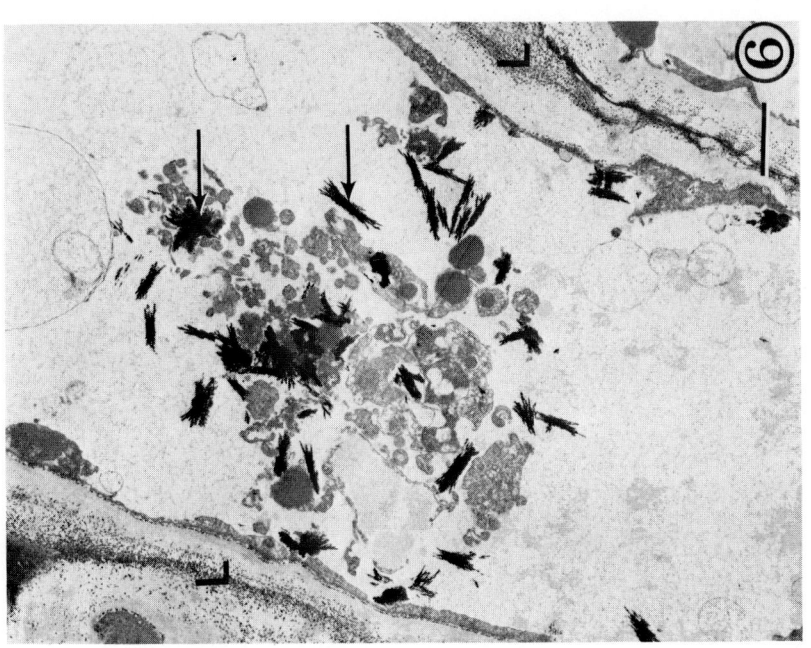

FIGURE 6. Pectinase reaction product (arrows) in vacuole and middle lamella (L) of mature laticifer. Bar = 2 μm.

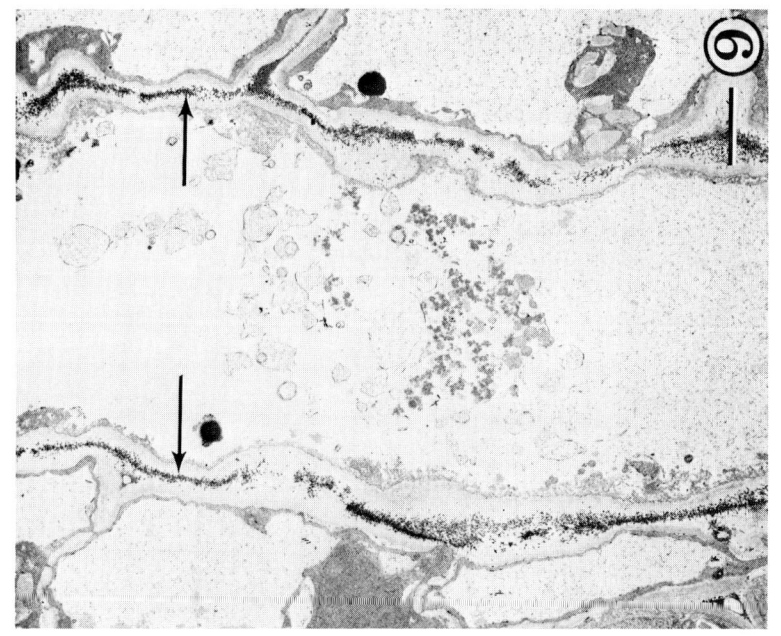

FIGURE 9. Laticifer of boiled control specimen. Reaction product is absent from laticifer vacuole, but persists in cell wall (arrows). Bar = 2 μm.

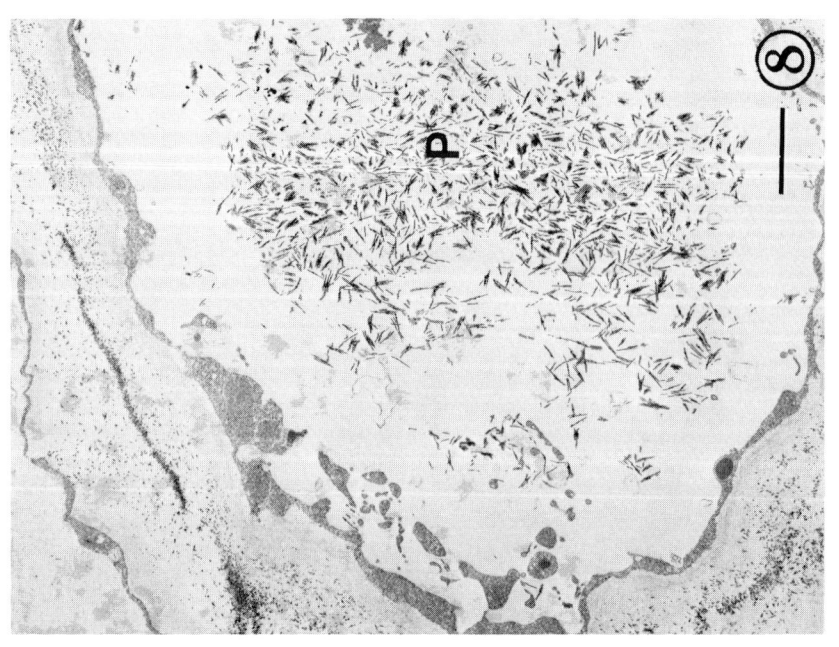

FIGURE 8. Diffuse pectinase reaction product (P) in laticifer vacuole. Bar = 2 μm.

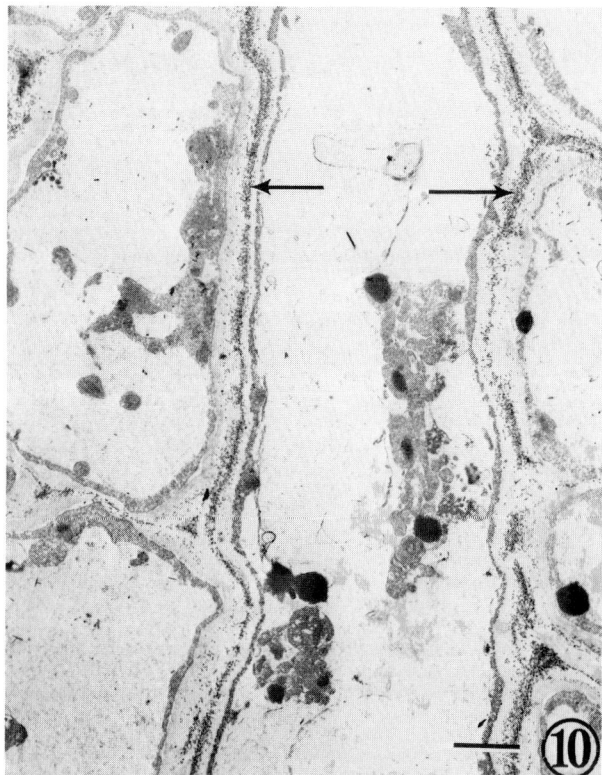

FIGURE 10. Laticifer of minus pectin control specimen. Note absence of vacuolar reaction product and presence of reaction product in cell wall (arrows). Bar = 2 μm.

REFERENCES

1. **Crookes, P. R. and Grierson, D.**, Ultrastructure of tomato fruit ripening and the role of polygalacturonase isoenzymes in cell wall degradation, *Plant Physiol.*, 72, 1088, 1983.
2. **Morré, D. J.**, Cell wall dissolution and enzyme secretion during leaf abscission, *Plant Physiol.*, 43, 1545, 1968.
3. **Rogers, B. J. and Hurley, C.**, Ethylene and the appearance of an albedo macerating factor in citrus, *J. Am. Soc. Hortic. Sci.*, 96, 811, 1971.
4. **Riov, J.**, A polygalacturonase from citrus leaf explants, *Plant Physiol.*, 53, 312, 1974.
5. **Greenberg, J., Goren, R., and Riov, J.**, The role of cellulase and polygalacturonase in abscission of young and mature shamouti orange fruits, *Acta Hortic.*, 34, 1, 1975.
6. **Jensen, W. A. and Fisher, D. B.**, Cotton embryogenesis in the tissues of the stigma and style and their relation to the pollen tube, *Planta*, 84, 97, 1969.
7. **Wilson, K. J., Nessler, C. L., and Mahlberg, P. G.**, Pectinase in *Asclepias* latex and its possible role in laticifer growth and development, *Am. J. Bot.*, 63, 1140, 1976.
8. **Allen, R. D. and Nessler, C. L.**, Cytochemical localization of pectinase activity in laticifers of *Nerium oleander* L., *Protoplasma*, 119, 74, 1984.
9. **Bal, A. K.**, Cellulase, in *Electron Microscopy of Enzymes: Principles and Methods*, Vol. 3, Hayat, M. A., Ed., Van Nostrand Reinhold, New York, 1974, chap. 3.
10. **Karnovsky, M. J.**, A formaldehyde-glutaraldehyde fixative of high osmolality for use in electron microscopy, *J. Cell Biol.*, 27, 137A, 1965.
11. **Spurr, A.**, A low-viscosity epoxy resin embedding medium for electron microscopy, *J. Ultrastruct. Res.*, 26, 31, 1969.
12. **Reynolds, E. S.**, The use of lead citrate at high pH as an electron opaque stain in electron microscopy, *J. Cell Biol.*, 17, 208, 1963.
13. **Mahlberg, P. G.**, Development of the non-articulated laticifer in proliferated embryos of *Euphorbia marginata* Pursh., *Phytomorphology*, 9, 156, 1959.

POLYPHENOL OXIDASE

Kevin C. Vaughn

INTRODUCTION

Polyphenol oxidase (PPO; E.C. 1.10.3.1) is a copper-containing enzyme that catalyzes the *o*-hydroxylation of monophenols to *o*-diphenols as well as the oxidation of *o*-diphenols to quinones.[1] Although it was previously supposed that this enzyme was involved in hydroxylation of monophenols to diphenols in vivo, plants that are devoid of PPO activity still produce the same quantity and types of *o*-hydroxylated phenols as plants that have PPO activity.[2] In fact, no known function for this enzyme has been established, despite its ubiquitous occurrence in the plant kingdom. Data from cytochemical[3-7] and subcellular fractionation[8,9] studies have uniformly established the plastid as the sole site of PPO activity in nonsenescent tissue. Virtually all types of plastid (leukoplasts,[3,7,10] etioplasts,[11] amyloplasts,[6] and chloroplasts[3-6,11]) have been shown to have PPO activity, although plastids of guard cells and bundle sheath cells are exceptions in that they lack PPO activity.[3,4]

BASIS OF THE CYTOCHEMICAL REACTION

Methods for light-microscopic cytochemical detection of PPO and the related enzyme tyrosinase have been available for many years.[12] The assay takes advantage of the fact that the product of the oxidation of some *o*-diphenols to their *o*-diquinones is a colored molecule of relatively low solubility in aqueous buffers. Sites of product accumulation are detected simply by incubating fixed tissue pieces in a substrate such as DL-dihydroxyphenylalanine (DOPA); the brown DOPA-quinone reaction product is readily detectable at the light microscope level.

Okun et al.[13] modified the basic light cytochemical procedure into an electron microscopic procedure for tyrosinase by reacting the DOPA-incubated sections in osmium tetroxide to form an "osmium black", easily detectable at the electron microscopic level. Plant scientists were quick to take advantage of the development in electron microscopic cytochemistry, and, shortly after the publication of the results on tyrosinase in animal systems, two reports describing the cytochemical localization of PPO in plant tissue were published.[14,15] Since these early reports, only one major modification of the procedure developed by Czaninski and Catesson[6,14] has been adopted: the use of cacodylate buffer (instead of phosphate buffer) for the DOPA solution.[16] The DOPA solutions appear to be much more stable in cacodylate buffers and virtually no autooxidation of DOPA occurs during the long incubation period.

Olah and Mueller[7] indicated that 3,3'-diaminobenzidene (DAB) would be a useful substrate for PPO detection. These authors found that leukoplasts of carrot-oxidized DAB to an osmiophilic polymer at the same sites of accumulation as the DOPA-quinone reaction products and were sensitive to the same inhibitors. However, as indicated by Frederick (Chapter 1), DAB serves as an electron donor to a number of enzyme systems (e.g., catalase, peroxidase) and reacts with heme-containing proteins throughout the cell. Thus, although DAB may be a superior reagent to DOPA as far as ease of detection after reaction with osmium, it is not useful in establishing the sole cellular location of PPO.

PROCEDURES FOR LOCALIZATION OF PPO (from Vaughn and Duke)

1. Fix small (1 mm²) tissue pieces in 2.5% (v/v) glutaraldehyde in 0.10 M cacodylate or phosphate buffers (pH 7.2) for 60 to 90 min at 0 to 4°C.

2. Wash tissue in 5% (w/v) sucrose in 0.05 M buffer of that type used in the initial fixation, two changes, 30 min each.
3. Incubate the tissue in 5 mg/mℓ DL-DOPA in 0.10 M cacodylate buffer (pH 7.2) for 18 hr at 0 to 4°C. After the 18-hr incubation, allow the tissue to incubate for an additional 60 to 90 min at room temperature in a fresh DOPA solution.
4. Wash the tissue repeatedly (six times, 5 min each) in 5% (w/v) sucrose in 0.05 M cacodylate or phosphate buffers (pH 7.2) at 0 to 4°C.
5. Wash the tissue in two changes (15 min each) of 0.10 M cacodylate (pH 7.2) at 0 to 4°C. Check the extent of washing by reacting a drop of the last wash with a small drop of 1% (w/v) OsO_4. If the drop turns black immediately, continue washing until the drop from a wash no longer turns black.
6. Postfix the tissue in 1% (w/v) OsO_4 in 0.10 M cacodylate buffer for 1 hr at room temperature.
7. Wash tissue in distilled H_2O, dehydrate, embed, and section as per usual procedures. Although urany/acetate and lead citrate poststaining are not recommended, in heavily stained specimens considerably more structural detail may be revealed if poststaining is performed.

Several kinds of controls may be run. Addition of the PPO inhibitor diethyldithiocarbamate (DDC) at 100 mM to the preincubation sucrose washes and, if necessary, to the DOPA solution as well completely eliminates the staining of PPO.[4,6,7] Minus substrate controls are effective for some tissues, but, because of the large amounts of phenols found in the vacuole of most plant cells (and the relatively free movement of these phenols after fixation), the endogenous phenols may themselves act as a substrate for PPO and react to form an osmium black. Boiling the tissue for a minute to inactivate PPO has also been used as a control,[6,7] but, because nearly all proteins would be denatured with this treatment, it is not an adequate test for PPO specifically. Peroxidase found in the cytoplasm, vacuole, and walls will accept electrons from DOPA in the presence of endogenously available H_2O_2, giving rise to a false PPO reaction in these cellular sites. Addition of 1 mM formate or 1 mg/mℓ catalase to the incubation will slow the action of peroxidase.[6,7]

ANALYSIS OF RESULTS

PPO staining is limited to the thylakoids of chloroplasts that have been properly fixed and stained (Figure 1A). The lumen of the thylakoids are darkened throughout the chloroplast, indicating a rather even distribution of this chloroplast protein along the surface of the thylakoid. In leukoplasts or developing immature chloroplasts PPO activity is often associated with large membrane-bound protein bodies, as well as the rudimentary thylakoids or vesicles that are found in these plastids (Figure 1B). The DDC controls show none of the reaction of the untreated tissue (Figure 1C).

Recently, immunocytochemical procedures have been used to verify that PPO is localized in the thylakoids, as shown by the cytochemical procedures, and, at least in the case of *Vicia faba* PPO, is a single molecular weight form.[17]

USES OF THE PPO CYTOCHEMICAL STAIN

Because of the ease of the cytochemical reaction for PPO, it is surprising that not more studies have been undertaken that utilized this procedure to answer basic physiological questions. There have been a number of studies demonstrating the value of this procedure. Meuller and Beckman[10] utilized the procedure to detect PPO in various tissues of cotton, and to determine that the distribution of PPO-containing cells did not correlate with the cells

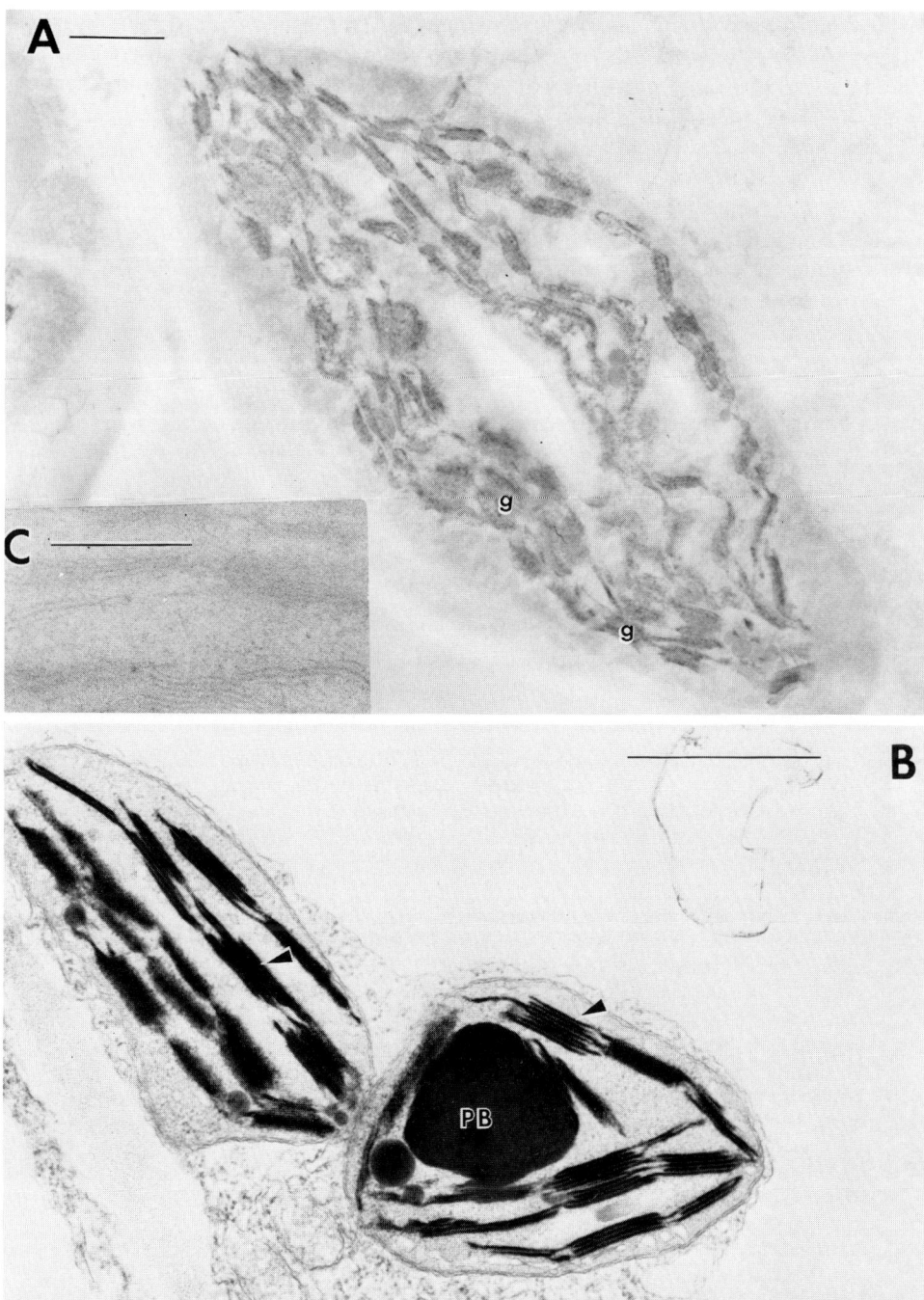

FIGURE 1. Cytochemical localization of PPO. (A) Mesophyll plastids of *Nicotiana* are strongly reacted in both the grana (g) and stroma lamellae. Compare the strong staining of the lamellae with the lack of staining of the envelope and other cellular structures; (B) Leukoplast from a developing cotton leaf contains a strong reaction in both the protein body (PB) as well as in the rudimentary thylakoids (arrows). Section was lightly poststained after reaction; (C) staining of the sections for PPO that have been preincubated in the PPO inhibitor DDC show no reaction along the thylakoids. Bar = 1.0 μm.

that had phenol-depositions in the vacuole. Martyn et al.[18] found that, in water hyacinth, infection of the plant with fungal pathogens leads to an increase in the PPO staining in the plastids. In our laboratory, the cytochemical stain has been heavily utilized in investigations of plastid development,[3-5] the effects of the fungal toxin tentoxin,[4,5] and the role of PPO in seed coat hardening.[19] From the results of our studies with the C_4 plant *Sorghum*,[4,5] it appears that PPO would represent an excellent marker enzyme for mesophyll plastids, because even at early developmental stages, mesophyll plastids possess PPO activity, while at all stages bundle sheath plastids are devoid of activity.[4,5] In studies of C_3-C_4 hybrids the segregation of this "trait" for exclusive mesophyll plastid occurrence of PPO may be an interesting one to study to see if this "trait" is inextricably linked to the C_4 syndrome.

REFERENCES

1. **Vaughn, K. C. and Duke, S. O.**, Functions of polyphenol oxidase in higher plants, *Physiol. Plant.*, 60, 106, 1984.
2. **Duke, S. O. and Vaughn, K. C.**, Lack of involvement of polyphenol oxidase in *ortho*-hydroxylation of phenolic compounds in mung bean seedlings, *Physiol. Plant.*, 54, 381, 1982.
3. **Vaughn, K. C., Miller, P. O., and Wilson, K. G.**, Ultrastructural localization of polyphenoloxidase in *Aegopodium podograria*, *Cytobios*, 31, 27, 1980.
4. **Vaughn, K. C. and Duke, S. O.**, Tissue localization of polyphenol oxidase in *Sorghum*, *Protoplasma*, 108, 319, 1981.
5. **Vaughn, K. C. and Duke, S. O.**, Tentoxin effects on *Sorghum*: the role of polyphenol oxidase, *Protoplasma*, 110, 48, 1982.
6. **Czaninski, Y. and Catesson, A. M.**, Polyphenol oxidases (plants), in *Electron Microscopy of Enzymes*, Vol. 2, Hayat, M. A., Ed., Van Nostrand Reinhold, New York, 1974, 66.
7. **Olah, A. F. and Mueller, W. C.**, Ultrastructural localization of oxidative and peroxidative activities in carrot suspension cell culture, *Protoplasma*, 106, 231, 1981.
8. **Arnon, D. I.**, Copper enzymes in isolated chloroplasts. Polyphenol oxidase in *Beta vulgaris*, *Plant Physiol.*, 24, 1, 1949.
9. **Henry, E. W., DePoore, J. M., O'Connor, M. N., and DeMorrow, J. M.**, Sorbitol-disrupted spinach (*Spinacia oleracea* L.) chloroplasts: cytochemical localization of polyphenol oxidase in discontinuous sucrose density gradient fractions, *J. Microsc. Cytol.*, 13, 365, 1981.
10. **Mueller, W. C. and Beckman, C. H.**, Ultrastructural localization of polyphenol oxidase and peroxidase in roots and hypocotyls of cotton seedlings, *Can. J. Bot.*, 56, 1579, 1978.
11. **Vaughn, K. C. and Duke, S. O.**, Tentoxin-induced loss of plastidic polyphenol oxidase, *Physiol. Plant.*, 53, 421, 1981.
12. **VanFleet, D. S.**, Histochemical localization of enzymes in vascular plants, *Bot. Rev.*, 18, 354, 1952.
13. **Okun, M. R., Edlestein, L. M., Or, N., Hanada, G., Donnellan, B., and Lever, W. F.**, Histochemical differentiation of peroxidase mediated and tyrosinase-mediated melanin formation in mammalian tissues, *Histochemine*, 23, 295, 1970.
14. **Czaninski, Y. and Catesson, A. M.**, Localization ultrastructurale dactivites polyphenoloxydasiques dan les chloroplastes de *Nicotiana glutinosa*, *J. Microsc.*, 15, 409, 1972.
15. **Parish, R. W.**, The intracellular location of phenoloxidases, peroxidase and phosphatases in the leaves of spinach beet (*Beta vulgaris* L. subspecies *Vulgaris*), *Z. Pflanzenphysiol.*, 66, 176, 1972.
16. **Eppig, J. J.**, Tyrosinase, in *Electron Microscopy of Enzymes*, Vol. 2, Hayat, M. A., Ed., Van Nostrand Reinhold, New York, 1974, 79.
17. **Vaughn, K. C. and Duke, S. O.**, Tentoxin stops the processing of polyphenol oxidase into an active protein, *Physiol. Plant.*, 60, 257, 1984.
18. **Martyn, R. D., Samuelson, D. A., and Freeman, T. E.**, Ultrastructural localization of polyphenoloxidase activity in leaves of healthy and diseased water hyacinth, *Phytopathology*, 69, 1278, 1979.
19. **Egley, G. H., Paul, R. N., Vaughn, K. C., and Duke, S. O.**, Role of peroxidase in the development of water-impermeable seed coats in *Sida spinosa* L., *Planta*, 157, 224, 1983.

Index

INDEX

A

AAT, see Aspartate aminotransferase
Acetate, 11
Acetone, 116
Acetyl choline esterase, 68
Acetyl-CoA, 135, 136, 142
Acid lipases, see also specific types, 125, 127
Acid phosphatases, see also specific types, 37, 47—49
Acrolein, 43
Adenosine diphosphate (ADP), 77
Adenosine monophosphate (AMP), 55
Adenosine-3'-monophosphate (3'-AMP), 112
Adenosine-5'-monophosphate (5'-AMP), 112
Adenosine triphosphatase (ATPase), 50
Adenosine triphosphate (ATP), 50
Adenylate cyclase, 55—56
5'-Adenylylimidodiphosphate (AMP-PNP), 55
ADP, see Adenosine diphosphate
Agarose, 74
D-Alanine, 25
Alcohol dehydrogenase, 76, 85
Aldehyde fixation
  AAT and, 96—97
  alkaline phosphatases and, 58
  cerium and, 25
  concentration of, 40
  controls and, 84
  dehydrogenases and, 73, 80, 83
  IDPase and, 52
  lipases and, 124
  malate synthase and, 136—138
  nuclease and, 109, 115, 119
  pectinase and, 149
  peroxidases and, 11
  phosphatases and, 38, 45
  polyphenol oxidase and, 159
  TPPase and, 53
Algae, see also specific types, 10—11
Alkaline lipases, see also specific types, 124, 125, 127
Alkaline phosphatases, see also specific types, 58—59, 124
Allium sp., 40
Amaranthus hybridus, 99, 100
D-Amino acid oxidase, 25
2-Amino-4-methoxy-3-butenoic acid, 98
2-Aminooxyacetic acid, 98
3-Amino-1,2,4-triazole (AT), 7, 12, 17, 25, 26
AMP, see Adenosine monophosphate
3'-AMP, see Adenosine-3'-monophosphate
5'-AMP, see Adenosine-5'-monophosphate
AMP-PNP, see 5'-Adenylylimidodiphosphate
Animal cell lipases, 123
Antibodies, see also specific types, 114
Artifacts, 17—18
Arum maculatum, 8

Ascaris suum, 133
Asclepias syriaca, 152
Ascorbate oxidase, 86
d-Aspartate, 98
l-Aspartate, 96—98
Aspartate aminotransferase (AAT), 95—104
AT, see 3-Amino-1,2,4-triazole
ATP, see Adenosine triphosphate
ATPase, see Adenosine triphosphatase
Atriplex sp., 97
Autolysis, 49
Autophagy, 49
Azide, 11, 12, 16
Azo dye, 37

B

BAXD, see *N,N'*Bis (4-amino-phenyl)-1,3-xylylendiamine
Bean, 95, 109, 133
BED, see *N,N'*Bis-(4-amino-phenyl)-*N,N'*-dimethyl ethylenediamine
Beef liver catalase, 7
Benedict's solution, 149, 151, 152
Benzidine, 3
Benzoin methylether, 116
Beta vulgaris, 58
Betula pubescens, 78
Binding of DAB, 17
*N,N'*Bis-(4-amino-phenyl)-*N,N'*-dimethyl ethylenediamine (BED), 19
*N,N'*Bis (4-amino-phenyl)-1,3-xylylendiamine (BAXD), 19
Buffers, see also specific types, 6
Bushbean, 95
Butyl methacrylate, 116

C

Cacodylate, 11
cAMP, see 3',5'-Cyclic adenosine monophosphate
Candida albicans, 124
Carbon dioxide, 29, 87
Carbon monoxide, 16
Castor bean, 109, 133
Catalase, 5—11, 15
Cauliflower, 95
C_3-C_4 hybrids, 162
Cell constituent isolation, 19
Cellular autolysis, 49
Cerium, 25—35, 112
Cerium chloride, 25—28, 33—35
Cerium hydroxide, 26
Cerium perhydroxide, 25
Chelated copper, 84
Chelators, 69—70

Chlamydomonas reinhardi, 10, 83
Chlorella sp., 69
Chlorogonium elongatum, 10
p-Chloromercuribenzoate (PCMB), 82, 86
Chlorophytum comosum, 109
Chloroplast electron transport, 29
Citrate, 69
CMP, see Cytidine monophosphate
CoA-SH, 134—136
Cobalt, 18
Cobalt chloride, 18
Coleus sp., 12
Colloidal gold, 117
Controls
    AAT and, 98
    for dehydrogenases, 83—88
    inhibitor, 85—88
    pectinase and, 151
    phosphatases and, 42—44
    polyphenol oxidase and, 160
Conventional transmission electron microscopy (CTEM), 25, 26, 32
Cooper, 69, 70, 142
    controls and, 84
    G-6-P and, 58
Copper oxide, 149, 151
Copper sulfate, 70, 80
Coprinus cinereus, 77
Cotton, 95, 160
CTEM, see Conventional transmission electron microscopy
Cucumber, 136, 138
Cucurbita sp., 111
Cyanide, 14
3′,5′-Cyclic adenosine monophosphate (cAMP), 55
Cynodon, 100
    dactylon, 98, 99
Cytidine monophosphate (CMP), 47, 58
Cytidylic acid, 54
Cytochrome c peroxidase, 14—15
Cytochrome oxidase, 13—16
Cytosol, 8

D

DAB, see 3,3′-Diaminobenzidene
DDC, see Diethyldithiocarbamate
Dehydrogenases, see also specific types, 65—88
    controls for, 83—88
    coupling agent, chelator, 69—70
    DMSO and, 74
    electron acceptors, 66—69
    electron microscopy, 77—83
    ferricyanide and, 68—69
    fixation and, 73, 78
    freezing and, 74
    incubation medium, 70—73
    inhibitor controls and, 85—88
    light microscopy and, 75—77
    NBT vs., 67
    postincubation and, 75
    preincubation and, 73—74
    substrates for, 70—71
    TNBT vs., 67
Dextran, 45, 56
3,3′-Diaminobenzidene (DAB), 3—19
    artifacts and, 17—18
    benzidine vs., 3
    binding of, 17
    cell constituent isolation and, 19
    cytochrome oxidase and, 13—16
    in cytosol, 8
    diffusion of, 17
    electron microscopy and, 17
    glyoxysomes and, 7
    hydrogen peroxide and, 5—8, 11, 12
    light microscopy and, 16—17
    microspectrophotometry and, 17
    organelles and, 16—17
    oxidation of, 4—5
    photosystem I and, 16
    polymerization of, 8
    polyphenol oxidase and, 16, 159
3,3′-Diaminobenzidene (DAB) tetrahydrochloride, 18
Dianthus sp., 77
Diazonium salts, 123
Diethyldithiocarbamate (DDC), 160
Diffusion of DAB, 17
Digestion, tryptic, 43—44
Dihydroxyphenylalanine (DOPA), 16, 160, 169
Dimethoxysulfoxide (DMSO), 27, 28, 72—73
    dehydrogenases and, 74
    lipases and, 123
    malate synthase and, 138, 140
Dimethylacetamide, 123
Dimethylthyazolyl tetrazolium bromide (MTT), 66
Disruption and dehydrogenases, 83—84
DMSO, see Dimethoxysulfoxide
DNase, 111, 112
DOPA, see Dihydroxyphenylalanine
Dryopteris
    filix-mas, 79
    sp., 85
Dye, see also specific types
    azo, 37
Dye-coupling methods, 113—114

E

EDTA, 76, 77
Electron acceptors, 66—70
Electron microscopy, see also specific types
    azo dye and, 37
    DAB and, 17
    dehydrogenases and, 66
    high-voltage (HVEM), 25, 26, 33
    lead phosphate and, 37
    lead sulfide and, 37
    lipases and, 123

NBT and, 67
   polyphenol oxidase and, 159
Electron transfer, 29, 73
Endo-exonucleases, 107
Endonucleases, 107
Escherichia coli, 133
Ethanol, 76
N-Ethylmaleimide (NEM), 82, 86
Exogenous phosphatase, see also specific types, 109—112
Exonucleases, 107, 108, 113

F

Ferricyanide, 70
   Chlorella sp. and, 69
   dehydrogenases and, 65, 66, 68—69, 73, 80
   malate synthase and, 134—135, 138, 139, 142
   NBT vs., 68
   pea and, 69
Ferricyanide oxidoreductase, 66
Ferricyanide reductase, 79—82, 84
Fixation, see also specific types, 38—41
   aldehyde, see Aldehyde fixation
   carbon dioxide, 87
   concentration of, 40
   dehydrogenases and, 78
   duration of, 39
   β-glycerophosphatase and, 39
   phosphatases and, 38
   post, see Postfixation
Fractionation
   dehydrogenases and, 83—84
   nuclease and, 109
Freezing and dehydrogenases, 74
Fructose-1,6-diphosphate, 58
Fructose-6-phosphate, 58
Fungi, see also specific types, 10—11

G

GAPDH, see Glyceraldehyde 3-phosphate dehydrogenase
Gelatine, 74
Germination, 49
Glucose-6-phosphatase (G-6-P), 57—58
Glucose-1-phosphate, 58
Glucose-6-phosphate dehydrogenase, 76
L-Glutamate, 77
Glutamate-aspartate transaminase, see Aspartate aminotransferase (AAT)
Glutamate decarboxylase, 87
Glutamate dehydrogenase, 77
Glutamate oxalacetate transaminase, see Aspartate aminotransferase (AAT)
Glyceraldehyde-3-phosphate dehydrogenase (GAPDH), 65, 76
Glycerin, 123
β-Glycerophosphatase, 39, 47

β-Glycerophosphate, 47, 49, 54, 58
DL-Glycerophosphate, 77
Glycine, 6, 76
Glycine sp., 28, 30
Glycolate dehydrogenase, 83
Glycolate oxidase, 26, 27, 33
Glyoxylate, 133, 135
Glyoxysomes, 7
Gold, 18, 117—119
Gold chloride, 18
Gomori reaction, 37
G-6-P, see Glucose-6-phosphatase

H

Hansenula polymorpha, 25
Hatchett's brown, 69
Heme compounds, 12
Heme protein activity, 16
High-voltage electron microscopy (HVEM), 25, 26, 33
Hordeum vulgare, 79
Horseradish peroxidase, 19
HVEM, see High-voltage electron microscopy
Hydrogen peroxide, 29
   catalase and, 15
   DAB and, 5—8, 11, 12
α-Hydroxy acid oxidase, 25

I

IDPase, see Inosine diphosphatase
Imidazole, 97
Immunocytochemistry of plant nucleases, 114—119
Immunoelectrophoresis, 119
Immunofluorescence, 115
Immunoglobulin G, 115—117
Immunoperoxidase, 115
Incubation
   duration of, 6, 71—72
   lipases and, 124—126
   pectinase and, 150
   phosphatases and, 45
   post, 75
   pre, see Preincubation
   temperature of, 6, 71—72, 125
   times for, 45
Incubation media, see also specific types, 70—73
   AAT and, 97—98
   pH of, 71
   phosphatases and, 42
Indirect immunofluorescence, 115
Inhibitor controls, 85—88
Inosine diphosphatase (IDPase), 50, 52, 53
Intercellular transport, 48—49
Ipomoea sp., 114, 115
Iron, 58
DL-Isocitrate, 77
Isocitrate dehydrogenase, 76—77
Isolation of cell constituents, 19

K

KCN, see Potassium cyanide
α-Ketoglutarate, 96—98

L

DL-Lactate, 77
Lactate dehydrogenases, 72
Latrix sp., 77
Lead
  AAT and, 96, 100
  alkaline phosphatases and, 58
  ATPase and, 50
  IDPase and, 50
  lipases and, 124, 126
  nuclease and, 112
Lead citrate, 142
Lead nitrate, 44, 45, 50, 56, 59
Lead oxalacetate, 97
Lead phosphate, 37, 44
Lead precipitation method, 109—112
Lead sulfide, 37
Leukocytes, 25
Light microscopy
  DAB and, 16—17
  dehydrogenases and, 66, 75—77
  malate synthase and, 143
  polyphenol oxidase and, 159
Lignified xylem, 12
Lipases, see also specific types, 123—130
  acid, see Acid lipases
  alkaline, see Alkaline lipases
Lupinus, 15
  luteus, 78
Lysolecithinase (lysophospholipase), 124
Lysophospholipase (lysolecithinase), 124

M

Magnesium, 58
Magnesium chloride, 6, 59
Magnesium phosphate, 142
Magnesium sulfate, 50, 56
Maize, 7, 48, 54, 56
DL-Malate, 80
Malate dehydrogenase (MDH), 65, 72, 77, 79—82
L-Malate glyoxylate-lyase, see Malate synthase
Malate synthase, 133—146
  aldehyde fixation and, 136—138
  DMSO and, 138, 140
  ferricyanide and, 134—135, 138, 139, 142
  light microscopy and, 143
  potassium ferricyanide and, 139
Malic acid, 70
L-Malic acid, 77
Malonate, 85
MAO, see Monoamine oxidase

MB, see Meldola blue
MDH, see Malate dehydrogenase
Meldola blue (MB), 72
Metal capture method, 112—113
Methanol, 25
Methanol oxidase, 25
1-Methoxy PMS, 72
Microspectrophotometry, 17
Monoamine oxidase (MAO), 65, 73
MTT, see Dimethylthyazolyl tetrazolium bromide

N

NAD, 76—77, 82, 85
NADH, 65, 70—72
NADH-diaphorase, 77
NADH-ferricyanide oxidoreductase, 80, 87
NADH-ferricyanide reductases, 68, 85
NADH oxidoreductase, 68, 69
NADH-tetrazolium oxidoreductase, 77
NAD-malic enzyme (NAD-ME), 95
NAD-ME, see NAD-malic enzyme
NADP, 76, 77, 82
NADPH, 65, 70—72, 77
NADPH-diaphorase, 77
NADPH oxidoreductases, 69
$NADP^+$ malic enzyme (NADP-ME), 95
NADP-ME, see $NADP^+$ malic enzyme
Naphthol, 37
Naphthol AS-BI phosphate, 37, 49
Naphthyl phosphates, 37
α-Naphthyl thymidine 3'-phosphate, 107, 113
α-Naphthyl thymidine 5'-phosphate, 107, 113
NBT, see Nitro-blue tetrazolium chloride
NEM, see *N*-Ethylmaleimide
Nerium oleander, 149, 152
Neutral lipase, 127
Nickel, 18
Nickel ammonium sulfate, 18
Nitella flexilis, 10
Nitrate reductase, 87
Nitro-blue tetrazolium chloride (NBT), 66—68, 75, 77
p-Nitrophenylphosphate, 47, 49
Nucleases, see also specific types, 107—119
  aldehyde fixation and, 115, 119
  exogenous phosphatase and, 109—112
  immunocytochemistry of, 114—119
  lead precipitation method and, 109—112
  plant, 114—119
  substrate for, 109
  terminology of, 107—109
Nucleoside phosphatase, 40
3'-Nucleotidase, 112—113

O

Oat, 95
Organelles, 16—17

Orinase, 58
Osmium, 18, 38
Osmium black, 3
Osmium tetroxide, 3, 17, 18, 160
Osmolarity, 71
Oxidation of DAB, 4—5

P

PAGE, 114
Palmitoyl-CoA, 25
Paraformaldehyde, 73
PCK, see Phosphoenol pyruvate carboxykinase
PCMB, see *p*-Chloromercuribenzoate
Pea, 69, 95
Pectinase, 149—157
Pediastrum tetras, 10
Penicillium sp., 117
Permanganate, 38
Peroxidase, 11—13, 160
Phenazine methosulfate (PMS), 72, 77, 85
p-Phenylenediamine, 19, 75
pH of incubation media, 71
Phleum pratense, 77
Phosphatases, see also specific types, 37—62
   acid, see Acid phosphatase
   adenylate cyclase, 55—56
   alkaline, see Alkaline phosphatases
   ATPase, see Adenosine triphosphatase
   azo dye and, 37
   controls and, 42—44
   exogenous, 109—112
   fixation and, 38—40
   glucose-6-phosphatase, 57—58
   IDPase, 52—53
   incubation media and, 42
   lead-based, 37
   localization, 45
   TPPase, 53—54
Phosphodiesterases, 107, 112—114
Phosphoenol pyruvate carboxykinase (PCK), 95
Phospholipases, 124
Photosynthesis, 95
Photosystem I, 16
Pinus lambertiana, 76
PMS, see Phenazine methosulfate
POCC, see Potassium-osmium-cyanide complex
Poly-[1,4-α-D-galacturonide glycanohydrolase], see Pectinase
Polymerization of DAB, 8
Polymorphonuclear leukocytes, 25
Polyphenol oxidase (PPO), 159—162
Polyvinylalcohol (PVA), 74
Polyvinylpyrrolidone, 74
Postfixation
   AAT and, 98
   lipases and, 126
Postincubation, 75
Potassium cyanide (KCN), 7, 11, 12, 16
   controls and, 86

   dehydrogenases and, 82
Potassium ferricyanide, 71, 135, 139
Potassium-osmium-cyanide complex (POCC), 18
Potassium phosphate, 140
Potato, 111
PPO, see Polyphenol oxidase
Preincubation
   dehydrogenases and, 73—74
   in ferricyanide, 138
   malate synthase and, 138
Propanediol, 6
Protease, 116
Protein, see also specific types
   heme, 16
Protein A-gold, 118, 119
Psychotria sp., 26, 30
Pteridium aquilinum, 79
PVA, see Polyvinylalcohol
Pyrocatechol, 19

Q

Quinine hydrochloride, 127

R

Redox potential, 65
Redox reactions, 66—69
Ribose-5-phosphate, 58
RNases, 107, 109, 111, 114, 115

S

Scanning electron microscopy (SEM), 25, 26, 30, 33
SDH, see Succinate dehydrogenase
Sedum telephium, 82
Seed germination, 49
SEM, see Scanning electron microscopy
Semicarbazid, 76
Semipermeable membranes, 74
Signal/noise ratio, 30
Sodium bisulfate, 76
Sodium cacodylate, 26
Sodium fluoride, 43
Sodium glycolate, 25
Sodium hypochlorite, 19
Sodium DL-lactate, 25
Sodium malonate, 79
Sodium pyrophosphate, 76
Sodium succinate, 76
Sodium taurocholate, 127
Sodium urate, 25, 26
Sonication, 74
Sorghum sp., 162
Soybean, 138
Spinach, 95

Substrates, see also specific types
    for dehydrogenases, 70—71
    for lipases, 126
    for nuclease, 109
    for pectinase, 150
Succinate dehydrogenase (SDH), 65, 69, 78—82
    controls and, 85, 87
    dehydrogenases and, 70, 74
    meldola blue and, 72
    NBT and, 75
Sucrose, 74, 126, 160
Superoxide radicals, 29

T

Tartrate, 69, 70, 80, 136, 139
TCA, see Tricarboxylic acid
Tellurite, 65
TEM, see Transmission electron microscopy
Temperature of incubation, 6, 71—72, 125
Tentoxin, 162
3,3',4,4'-Tetraaminobiphenyl, see 3,3'-Diaminobenzidene (DAB)
Tetrahydrochloride, 18
Tetramethylbenzidine (TMB), 19
Tetra-nitro BT (TNBT), 66, 67, 77
Tetrazoles, 66
Tetrazolium reductases, 85
Tetrazolium salts, 65
Theanol, 77
Thiamine pyrophosphatase (TPPase), 42, 53—54
2-Thiolnonanoylbenzanilide, 123
Thylakoids, 30
Tissue blocks, 7
TMB, see Tetramethylbenzidine
TNBT, see Tetra-nitro BT
Toad, 138
Tobacco, 109
TPPase, see Thiamine pyrophosphatase
TRAB, see Triethanol amine-HCl-buffer
Tradescantia sp., 112, 114, 115, 117, 119
Transmission electron microscopy (TEM), 25, 26, 30, 32
Transport
    ATPase and, 50
    chloroplast electron, 29
    intercellular, 48—49
Tricarboxylic acid (TCA) cycle, 133
Tricitum sp., 113

Triethanol amine-HCl-buffer (TRAB), 76, 77
Tris, 6
Tris-HCl, 11
Tris-maleate, 25, 26, 50
    adenylate cyclase and, 56
    alkaline phosphatases and, 59
    phosphatases and, 45
    TPPase and, 53
Tryptic digestion, 43—44
Tulip, 109
Turbatrix aceti, 133
Tyrosinase, 159

U

Uranyl acetate, 142
Urate oxidase, 26, 27, 33

V

Vicia, 111
    faba, 113, 160

W

Water hyacinth, 162
Wheat, 95, 111, 112
Wounding, 49

X

X-ray microanalysis (XRMA), 25, 30
    of cerium, 32
XRMA, see X-ray microanalysis
Xylem, 12

Y

Yucca sp., 26

Z

Zea mays, 7, 48, 54, 56
Zinc, 58